William Francis Ganong

The Teaching Botanist

A Manual of Information upon Botanical Instruction

William Francis Ganong

The Teaching Botanist
A Manual of Information upon Botanical Instruction

ISBN/EAN: 9783337165451

Printed in Europe, USA, Canada, Australia, Japan

Cover: Foto ©berggeist007 / pixelio.de

More available books at **www.hansebooks.com**

THE

TEACHING BOTANIST

A MANUAL OF INFORMATION UPON
BOTANICAL INSTRUCTION

TOGETHER WITH

OUTLINES AND DIRECTIONS FOR A COMPREHENSIVE
ELEMENTARY COURSE

BY

WILLIAM F. GANONG, Ph.D.

PROFESSOR OF BOTANY IN SMITH COLLEGE

New York
THE MACMILLAN COMPANY
LONDON: MACMILLAN AND CO., Ltd.
1899

PREFACE

It may appear at first sight that the title of this work is of wider scope than its contents. Addressed broadly to the teaching botanist, and professing to serve as a manual of information upon botanical instruction, it nevertheless deals chiefly with one phase of botanical teaching; namely, with its elementary presentation as a science. Yet in this it represents the actual condition of the problems in botanical teaching at the present day. On the one hand, our university and advanced college teaching, carried on as investigation or in that spirit, represents a well-nigh ideal relation of teacher to student; and on the other, the botanical part of nature study in the lower schools has hardly yet begun to attract the attention it deserves. It is just between these two, between advanced college and lower school work, that is, in elementary courses in Botany treated as a science, whether in high school or college, that the present problems lie. Here is the real centre of discussion, effort, and advance in botanical teaching to-day.

This part of botanical teaching is now in a state of wonderfully rapid expansion and transition. Three causes are contributing to this result: first, the natural reaction from its former extreme backwardness; second, a widening recognition of the value of the sciences, of which Botany is a leading one, in general education;

third, its acceptance as an entrance subject by some of
the leading colleges.   There is thus arising an unprece-
dented demand for a teaching that shall be more exten-
sive, more thorough, and more representative of the
present state of the science.   But so rapid is the advance in the science itself, as well as in methods of
teaching it, that only specialists with the best opportunities are able to keep pace with their progress, and
others are placed at a great disadvantage, no matter
how great their desire to improve their teaching or
their eagerness to utilize the latest and best knowledge.
It is for teachers of such progressive spirit, and in the
effort to bring together the best that is at present known
about the teaching of elementary courses in the science
of Botany, that this book has been written and made as
nearly as possible monographic in its character.   That
it may meet the rising demand for guidance in the
teaching of Botany in its more modern and approved
aspects, and even serve in some measure as a stimulus
to yet greater advancement, is the aim and wish of its
author.

The elementary course recommended and outlined in
this book, while it seems to me the one best suited to
conditions now prevailing and likely to prevail for a
long time to come, is certainly very far from final.   It
contains many topics that belong in the lower grades,
topics which are already there in some schools and
will soon be placed there in others.   But in the present
confused state of botanical teaching in the schools, it
is unsafe to assume in any general course that any
particular subjects have already been studied by the
students; and the only logical plan is to start the
course from the beginning, and then make special allow-

ances for those who have had particular topics. Again, although I have tried to utilize the best in the experience of my fellow-teachers (including the very valuable Report of the Committee of Ten) as well as in my own, yet the subject is so new that much discussion, trial, and experiment will be needed before the best selection and proportioning of subjects will be found.

It will be noticed that the plan of the course does not aim directly at what most of our leading teachers now regard as the ideal. This ideal places vital phenomena first, especially as they manifest themselves in moulding the physiognomy of vegetation. Indeed, I suppose most people will agree on this point, that the best knowledge we can give our students is that which will enable them to understand the influences determining the physiognomy or topography of vegetation, why it has the shapes and colors and sizes and distributions it has. Our courses should aim to do for the students of vegetation much the same that the modern science of physiography is doing for students of the forms of the land, — it should demonstrate the factors determining its present construction. But a course aiming directly at this ideal is not at present practicable, and indeed I doubt if it ever will be. The problem of the topography of vegetation is enormously more complex than that of the topography of the land, and we have not yet either the knowledge, the methods, or the inclination to attack it directly in general courses. It should be kept in view as an ideal and worked toward indirectly. The best basis or skeleton for an elementary course is, and I think always will be, not phenomena, which are abstract and elusive, but structure, which is concrete, and

with which the student may be brought into immediate
personal contact.   Using it as the basis of a course, as I
have done, has moreover this further advantage, that it
enables us to keep and build upon what is best in our
botanical courses of the past and present, and that best
has always been structural.

In the preparation of this work, I have made constant
use of these modern and excellent works, often cited in
the following pages, by Spalding, Bergen, Setchell, L. H.
Bailey, Barnes, and Atkinson, and I express here my
indebtedness to them.   The figures are all new, though
Nos. 17, 18, 22, 23, 24, have already appeared in an article of mine in the *Botanical Gazette* for April, 1899.
At least one valuable feature in the course given in this
work, namely, the use of the vertical diagrams of flowers, I owe directly to Professor Goodale ; but there is
much else in the book, particularly in its general spirit,
which I owe to his teaching.   I wish also to express
my obligation to Mr. E. J. Canning, Head Gardener
at Smith College, for his assistance in seeking new
materials for study and in devising easier ways of growing them, and to my assistant, Miss Grace Smith, for
her advice and aid in the proof-reading.

CONTENTS

VII

PART II

*AN OUTLINE FOR A SYNTHETIC ELEMENTARY
COURSE IN THE SCIENCE OF BOTANY*

DIVISION I

THE PRINCIPLES OF THE SCIENCE OF BOTANY

DIVISION II

THE NATURAL HISTORY OF THE GROUPS OF PLANTS

INTRODUCTION

It is plain to all who read the signs of the times that we in these days are seeing the beginning of a great movement toward the introduction of the Sciences into general Education. The chief obstacle to their rapid utilization, however, next to the conservatism of educational bodies, is the widespread ignorance of how to use them economically. In their modern form they are too new, and their advance has been too rapid, to let them grow gradually into the framework of our educational system, and they are forcing themselves in upon it from the outside. It becomes, therefore, a problem, and a difficult one, how best to assimilate them with what is already there, and as well how to present them in their optimum value and with the greatest possible economy. This book is an attempt to face squarely these issues for the Science of Botany.

From the point of view of general Education, the subject is not so vast as it seems. It has nothing to do with the special study of particular phases of the Science as followed in the universities, and only indirectly anything to do with that use of Plants and their phenomena which, as a part of Nature Study, is coming

to form so valuable an element in the discipline of the
primary schools, and so excellent a preparation for the
proper study of the subjects as Sciences later.   It is
limited rather to that systematic, complete, well-proportioned presentation of the salient facts and principles
of Botany essential to its treatment as a Science.   Such
a treatment the best colleges aim to give in their general or elementary courses, and its equivalent they are
beginning to expect from the preparatory schools as
an optional entrance requirement.   The time, therefore,
has come for opening up the question, What is the
Optimum of training and knowledge in an ideal elementary course in the Science of Botany, and how
may it most economically be realized?

At the outset it may be thought of little use to
attempt to point out an optimum treatment for this or
any Science, since the Optimum must necessarily be
subjective and vary with place and person, and since,
also, various practical reasons, such as imperfect training of teachers, different standards of colleges and
communities, lack of proper facilities, must oftener than
not prevent the attainment of any high standard.   To
this it may be answered that the practical difficulties
are constantly and rapidly lessening; that colleges and
communities are tending to become more uniform in
the essentials of their requirements; that, moreover,
an objective Optimum for the treatment of this as of
any Science, difficult though it may be to establish it,

must, in the nature of the case, exist; and that even
where the attainment of the Optimum cannot imme-
diately be hoped for, it is, nevertheless, an immense
advantage to have it ever in view as a standard of
comparison and an ideal for which to strive.

Naturally there will be much difference of opinion
as to what should enter into an optimum elementary
course in Botany treated as a Science, and it is only
after long discussion and much experiment that any
consensus of opinion can be expected. But all will
agree that the Optimum must be such as will make
available the potentialities of the Science as a means
for training the mind and as an element in culture; and
I think it will not be denied that to accomplish these
ends it must embody the essence of the best human
knowledge of the leading divisions of the Science, and
that it must include training in those qualities by which
that knowledge has been gained. The field of scientific
Botany should be laid open to the student, much as
the topography of a country is represented on a good
map; that is, however small its scale, it should show all
the great features in their proper relative proportions.
If this be admitted, it follows that the optimum course
in Botany must treat the science, not by divisions, but
synthetically, must include training in the elements of
Anatomy, Morphology, Physiology, Ecology, and Classification, and cannot be limited to any one or two of
them. The particular place and correlation of these I

shall try to develop in a later chapter. I wish here
merely to make plain the leading idea which is at the
foundation of this book.

The introduction of the principles of Physiology
and Ecology into elementary courses in the Science
of Botany is not only forced upon us by the present
state and trend of the Science, but is of immense pedagogic value as well. Plant structure, and relationships
as based upon structure, have been relatively so well
studied that investigators are coming to concern themselves more with the forces and influences which determine structure. Physiology is giving the interpretation
of Anatomy, and Ecology that of Morphology. The
conception of the Plant as first of all a living, breath-
ing, working being, with its functions controlling its
structure, is not only the truest, the most objective conception of it, but is as well the one which excites the
greatest human sympathy and interest. It is, then,
not only unfair to our students to continue to offer
them a conception of lesser worth, but also is a
refusal to accept the best "method" which the Science
has at present to give us. The introduction of Physiology and Ecology is the most marked characteristic
of progress in botanical teaching to-day, and amply
explains their prominence in the present work.

Those who have observed the rapid advances in the
teaching of the Biological Sciences in the past few years
must have noticed the answering changes in the books

devoted to it. First of all in time was the Text-book, the fount of all knowledge, descended to us from early times. The introduction of direct inductive study of facts and phenomena in the laboratory led, if not to the abandonment of the Text-book, at least to its temporary eclipse; and there rose to prominence the Laboratory Manuals, which were books of directions to enable the student economically to work out the more important truths for himself. But laboratory study alone, necessarily confined to a few types, has been found to give too disconnected a view of the Science, and the Text-book for supplementary reading is coming again into favor; and in some recent books we see an attempt to combine Laboratory Manual and Text-book. The Laboratory Manuals, also, have not proved altogether satisfactory, for they necessarily call for definite materials, which oftener than not are wanting when most needed; and, moreover, a good teacher will not submit to the restraint they impose upon him, especially in matters of detail, when they are placed in the hands of his students. To meet these difficulties the later books either give an excessive abundance of exercises, far more than can be accomplished in the assigned time, hence allowing a considerable choice, or else they are addressed less to the pupils than to the teacher, for whose benefit many pages of advice are added. This tendency to aim at influencing the teacher directly, to educate and advise him while leaving him free in his choice of methods

and materials, is most healthful, and directly in the
line which produces success in all occupations. As a
whole, the logical and desirable outcome of present
tendencies in laboratory instruction, so far as books are
concerned, seems to me this, — to separate the Laboratory Manual entirely from the Text-book, and to make
the former a series of model outlines for the use of the
teacher only, who, aided by suggestions as to alternative materials, etc., will make up from them for each
exercise special outlines to be given to the students,
which shall fit exactly the material at their disposal,
their state of advancement, and the particular mode of
treatment the teacher prefers; while the Text-book, used
only for supplementary reading after the laboratory
work, shall be truly a book of texts in the old sense,
as synthetically and attractively written as possible. It
is a Laboratory Manual of this type, addressed to the
teacher alone, with model outlines and advice upon
teaching the Science, which is offered by the present
book.

The leading idea in the construction of any set of
model outlines must, of course, be that of the Optimum,
an Optimum which is a resultant between practical conditions and their author's opinion as to what is the best
kind of a botanical course, both for training and knowledge. On this latter point opinions differ widely, and
I shall consider the leading ones in a later chapter; but
there is one which is so fundamental as to need men-

tion here. There are some teachers who believe that the first duty of the Sciences to-day is to supply the conspicuous lack of training in observation and inductive reasoning in the general educational system, and that those Sciences and those special parts of them best fitted for this purpose should be used, while other parts are more or less negligible. In other words, they believe that, at least for the present, the Sciences should be dependent organs of the general educational system, not independent members of it. There are books written with this idea more or less prominent. If the Sciences could yield little beyond mind-training, or if mind-training were the only aim of Education, this position would be sound. But the Sciences to-day are coming to mean much more in Education than the mere stopping of a gap in the general system, more than any certain kind of training, even more than a kind of knowledge made desirable by the activities of the times; they are coming to mean nothing less than a full and perfect equality with any and all other subjects whatsoever as elements in culture. This conception of the place of Botany in Education demands much more than the use of such parts of it as are particularly good for inductive training; it demands its treatment as an entity, complete and well-rounded, which implies, again, an objective Optimum as the ideal. I believe such an objective Optimum exists, though much experiment and discussion are needful before it will be found. Never-

theless, some of its characteristics are plain, and these
are embodied in the recommendations given in this
book, and particularly in the Outlines in Part II.

It is easy to have ideals, hard to attain to them.
It is true there are few teachers capable of using
such an optimum course; it is true that for many
schools it seems so vast in scope as to be unrealiza-
ble with actual conditions; and the materials, apparatus, and facilities appear hopelessly expensive. To
attempt to realize it everywhere at once would be
but to undertake an impossibility; but present-day
conditions are rapidly removing the chief obstacles.
In some colleges, teachers are being trained who can
teach these different divisions of the subject, and
teach them well. Further, by a rigorous selection of
the most fundamental topics, by a correct correlation
of these so that they may throw light upon one
another, by the invention of experiments which, with
simple appliances, logically demonstrate leading principles, by the finding out of those plants combining
the greatest illustrativeness with ease of acquisition,
by the use of the psychologically best methods, and
by other inventions aiming to secure greater efficiency
from the time, labor, and money expended; by all
these may a great advance be made toward the Optimum without proportionate cost.

Naturally, these economies must be worked out by
those who have the best facilities and talents for them;

and there is here opened up an attractive field for
investigation, one which is far broader than a search
for new methods, for it is a great study in correlation,
materials, and generalship. He who gives us a more
objective proportioning of subjects, a more remunera-
tive treatment of a topic, or a new device for the logical
proof of a fundamental principle, renders to Education
a service like his to humanity who makes two blades
of grass grow where one grew before. Great advances
in this direction are being made; to utilize these, and
even to add to them, has been a leading impulse in the
development of this work.

The Botanical course of the near future must be
more adaptive to Education, more broad and representative of the Science, more economical of energy than
in the past. It remains to inquire how these improvements may best be made.

PART I

ESSAYS ON BOTANICAL PEDAGOGICS

I. THE PLACE OF THE SCIENCES IN EDUCATION, AND OF BOTANY AMONG THE SCIENCES

IT is essential to the success of the teaching botanist that he have as clear-cut and objective a conception as possible of the place of his subject in Education. This he must work out for himself through observation and much thought, but here follow some data and ideas with a direct bearing upon it.

What is the aim of Education? Though so old, this question is yet ever new, and there is no subject of equal public importance still so little understood by those whom it most concerns. Its value in the abstract is everywhere granted, but there is still widely prevalent the greatest confusion between Education, Knowledge, Information, and Professional or Technical Training; and it is a first duty of every educator to acquire clear ideas upon these matters, and on all proper occasions vigorously to set them forth. Now the cosmic basis of Education seems to me this. Man is an animal whose weak and weaponless body is inferior to that of many of the brutes, but he has risen to domination over them, and much more of Nature besides, through the possession of one supreme char-

acteristic, Mind. To enable him not only to make
the best present use, but to realize the utmost potentialities of this his great weapon, — such is the true
end of Education. But though the aim is simple, its
attainment is hard, for Mind at its best is so complex, so liable to maladjustment, so little understood
by most of its possessors, that its cultivation requires
the greatest wisdom, skill, and sympathy, and the need
for these qualities is but partially realized by the public.

Of all of the many problems of Education, the one
nearest the surface is this, — to determine the proper
balance between mind-training pure and simple, and
the training of the mind for the practice of a particular business. The demand for the latter is incessant,
particularly in poorer communities; but if there is one
thing plain about Education, it is this, — that from
the very nature of Mind, it is a more efficient weapon
for any particular service if it has first been put into
a state of general sharpness and polish. This means
that each mind will achieve greater success in the
end, if, before it is turned to a particular work, it is
given the best general drawing-out it is capable of.
To give this is the first duty of Education.

Unhappily Education has to contend not only with
misunderstanding from without, but against dissensions
from within. In the abstract all admit that in mind-
training the leading faculties are to be drawn out,
and of these the following are of most account. First,

since all knowledge comes to us through the senses and by reasoning upon what they teach, training is necessary in the accurate use of these, and in drawing correct conclusions from their evidence, an inductive discipline best yielded by the Natural Sciences. Second, there is Number and its properties and reasoning thereon, largely deductive, which is the province of Mathematics. Third, there is Communication, involving expression; that is, Language. Fourth, there is relationship with other men, expressed in History and Political Economy. Now it would seem to be the duty of a good system of Education to give fair attention to all of these. But in fact, what is the case? As a fair index of what is generally regarded as important in Education to-day, we may take the entrance requirements of the large colleges. With few exceptions, these give the first and preponderating place to Languages, and among these it is not the one the student speaks that is made of most account, but two or three foreign ones. The second place is given to Mathematics. The third is given to a little History, which, however, is not of a kind to throw any light upon the constitution of affairs to-day, but is generally entirely ancient. Usually there is no fourth place, and if there is, it is given to a small amount of one of the Sciences timidly admitted as a partial alternative for one of the several Languages. Nor can the colleges claim that this does not repre-

sent their idea of what is important, but that it is
forced upon them by the poor teaching of the other
subjects in the preparatory schools; for many colleges
in their requirements for graduation, over which they
have full control, still demand much ancient Language
and the Mathematics, but leave the Sciences and History voluntary, thus stamping the former as essential,
and the latter as unessential to a good education.
The old requirements are still held to, and little note
is taken of the fact that the marvellous advance of the
Sciences in modern times has brought them into the
closest possible relations with every phase of human
life, and that the spread of democracy has created a
need for training in the knowledge essential to citizenship. Our system of Education is in many respects
a survival from ancient times, and in these days often
has less the appearance of an attempt to fit a student
for the conditions of modern life, than of an aim to
separate him as far as possible from the modern
world, and place him in a class based upon artificial
distinctions, a phenomenon only explainable as a per-
sistence among us of a tendency to fetich-worship.
Happily, these ancient ideals, though widely, are by
no means universally prevalent, but even the most lib-
eral and advanced universities have hardly yet thrown
them off altogether. It is little wonder that critics
are so often found to declare that much of our system contributes more to pedantry than to usefulness,

or that self-made men so often have no respect for it. These at least have had the clearness of sight to seek and use the inductive or natural method of acquiring their knowledge, a method vastly better than any which Education had to offer them; and there is many a man who has succeeded rather in spite of his education than in consequence of it. The self-made man has no doubt often been better made than would have been the case if " Education" had had anything to do with the process.

That this maladjustment of studies is as bad in practice as it is illogical in theory, every teacher of the Sciences knows. Through it, children, during the period of school study, when their minds are in the most receptive and formative state, so far from being made accustomed to natural inductive methods of acquiring knowledge, are subjected to excessive text-book and deductive work, which always tends to make them distrustful of their own powers, and leads them to regard as the only real sources of knowledge the thoughts of others properly recorded in printed books. The revival in students of the spirit of inductive inquiry, a spirit which they naturally possess, but which is usually crushed out of them by their school course, is the first and greatest task of any teacher of a Science. Against a system which permits such a condition, every teacher should wage determined and incessant battle.

The argument for a classical course has always

c

been that it conduces more than any other to the
attainment of Culture.   It assumes as true two things
now widely doubted, — first, that Culture can best be
obtained in the same way by all men, and second,
that the Sciences and History are inferior as cultural subjects to the Classics and Mathematics.   That
Culture is the first aim of Education, we all agree,
but in what does it consist?   Whatever it may have
been in the past, the conditions of modern society
are rapidly fixing this standard, that it does not consist in knowledge or training of any particular, pre-
determined kind, but rather in thorough knowledge
and training of some special useful kind, combined
with a general knowledge of what is passing in the
world around.   Culture consists less in wide knowledge than in wider sympathy; not so much in stores
of facts as in ability to transmute facts into knowledge; not only in well-grounded conviction, but in
toleration; not alone in absorption of wisdom, but as
well in its radiation; in patriotism that is without
provincialism; in the development of character.   But
since individual minds differ much in their composition, no one kind of treatment can be best for all,
and the ideal system will be that which is elastic
enough to allow each to receive what is best for it.
True culture, then, cannot be attained by forcing all
minds into any one mould, however carefully that
may be made, but is rather to be attained by allow-

ing each mind to expand for itself under a proper combination of nourishment from within and stimulus from without.

Whether or not the Sciences are inherently as efficient as the Classics in securing Culture is still in debate. The Sciences in these days have many distinguished advocates, of whom the greatest was Huxley, and, in this country, President Eliot. In the works of the former and the published addresses of the latter the interested reader may find the fullest satisfaction. As to knowledge, each must judge for himself whether it is not at least as conducive to gentle conduct, to good citizenship, and to sympathy with all grades of humanity, to know something of the forces with which man to-day is subduing Nature, of the processes going on in our own bodies, of the basis of the germ nature of disease, of what moves an electric car, of the meaning of the procession of the seasons with their manifold phenomena, as to know the lore and literature of Greece or the tongues of modern Europe, fine though these things be. And as to training, one may compare the best that can be claimed for the æsthetic value, the facility in expression, the polish of manner that the Languages help to produce, with the following description from Huxley of the training value of the Sciences.

"Science is, I believe, nothing but *trained and organised common sense*, differing from the latter only as the veteran may differ from a raw recruit: and its methods differ from those of common sense only so far as the guardsman's cut and thrust differ from the manner in which a savage wields his club. The primary power is the same in each case, and perhaps the untutored savage has the more brawny arm of the two. The *real* advantage lies in the point and polish of the swordsman's weapon; in the trained eye quick to spy out the weakness of the adversary; in the ready hand prompt to follow it on the instant. But, after all, the sword exercise is only the hewing and poking of the clubman developed and perfected.

"So, the vast results obtained by Science are won by no mystical faculties, by no mental processes, other than those which are practised by every one of us, in the humblest and meanest affairs of life." ("Science and Education," 1894, p. 45.)

To the equality of the Sciences with the older subjects in Education, a large and increasing part of the educational community is giving its assent. This finds its best expression not only in the curricula of our most progressive colleges, which put them on an equality in their own courses by making all subjects alike elective, but also in the newly-proposed Harvard entrance requirements, which, with restrictions made necessary by the imperfect way in which most of the Sciences are at present taught in the schools, require some one of them to be offered for entrance, while other colleges are coming to accept them as alternatives. This is the beginning of a movement whose logical end is the elevation of the Sciences to full educational rank with any and all other subjects. It must come about slowly, but it is the ideal which every teacher of the Sciences should never cease to strive for.

But how are the Natural Sciences to be admitted to full equality with subjects which not only are in possession of but fully occupy the ground? That the curriculum is already full there is of course no question, and the introduction of other subjects is possible only through either first, reducing the number of, or the time given to, those already there, or else second, by adding the new ones as alternatives to the older, permitting students to choose those from which they gain most good. The first is not to be thought of, for the older subjects, Languages and the Mathematics, are well entitled to the place they have won, and there are many students to whom they will mean more than the new subjects; their only offence is their unwarranted exclusiveness. The second is, I think, the logical, profitable, and inevitable solution, which will find its ultimate expression in the offering by schools, as colleges do now, of as complete, thorough, and extended courses in some one or more of the Sciences as they offer in Classics or Mathematics. This implies election in the school course, though not a free election such as some colleges find best, nor a possibility of ungoverned specialization, but election by groups, in which the class of subjects from which the student derives the greatest cultural good forms a major about which others, of a kind and in the proportion which experience shows to be most profitable, will be grouped. Thus will a student's greatest interest become the centre of his education and the point from which other subjects will be approached, to the great advantage of their learning.

To even a limited elective system in the schools, there are three objections often urged, which I shall here briefly examine: first, a student does not often know while at school what his particular bent is; second, most schools cannot offer so many courses, especially since those in the Sciences are more expensive than other kinds; third, early specialization is bad. As to the first, it is certainly true that few students find out their own specialties at school. But is not the discovery of this a chief function of the teacher? Surely the very first "method" in teaching is the diagnosis of each individual case and the fitting of its proper treatment to it. Every good teacher must notice what each of his students takes up most eagerly and learns most easily, and this knowledge may be made the basis of influence which will determine the student's election. Indeed, I think a very first duty of teachers is to find out what each student is best fitted for and to set him in the way of it. As to the second point, the expense of additional courses, it is true that scientific courses are expensive, though by no means to the degree popularly supposed. Without question, however, no system of Education, no matter how complete, will ever expect that any considerable number of the Sciences must be taught in every school, but will simply expect from the smaller the thorough teaching of one or two. It is the scientific method or spirit which is important, and for this, one Science is about as good as another. All advocates of the value of the Sciences in Education agree that one or two, well taught, are of far more worth than fragments of several; and the newer college entrance requirements offer a wide selection of subjects, but expect thorough preparation in the one or more selected. The least, however, the schools will be expected to offer will be a thorough course equal to the elementary courses in college, so that the student may there enter upon second courses as he now does in Languages and Mathematics.

It may seem an objection to this specialization in the Sciences that they are more or less interdependent, and one cannot be understood without some knowledge of facts and principles of the others; but this should be, and no doubt will be, met by continuous and ample elementary scientific study (of which

Nature Study is a part) in the lower schools, where
the principal facts, phenomena, and terminology, of
Physics, Chemistry, Botany, etc., will be learned at
first hand, thus forming the best possible basis for
the study of some one of those subjects as a Science
in the high school or of others in the college. Indeed, from all points of view, Nature Study, in order
to be really effective, needs to be thorough, considerable in amount, and unbroken from kindergarten to
high school. It is only thus that the natural inductive faculties of children can be preserved intact,
not to mention improved, through their school life.
Of course the principle of selection of certain Sciences to be taught in the smaller high schools will
in time lead to a similar selection of Languages,
whereby fewer of these being taught, the burdens of
the school will not be increased by the proper teaching of the Sciences. And this selection of fewer subjects for better teaching will be rendered the easier
because of the more uniform requirements toward which
the colleges are all tending.

The third objection, that early specialization is bad,
is based, as I think, upon a wrong idea of it. Specialization is by no means a selfish isolation in a nar-
row line of interests, but rather it consists in making
one's greatest interest the axis for the grouping of
the others. The conditions of modern life have set-
tled it for us that the only well-educated man is a

specialist, one who knows something well, it matters not so much what, and has sympathy for other things. Of this condition all Education should take account, as it is doing plainly enough in its higher grades. But even for the schools the value of the principle of specialization needs no argument, for it is there admitted even though its operation is confined to but one group of subjects, the Classics. Curiously enough, however, it is those who contend for the value of spending half of the school time through several of the school years upon the single subject of foreign languages who are quickest to condemn any such devotion to another group of subjects. I think their system for their own subjects is entirely correct, and it is simply the same privilege for the Sciences, and no more, that I ask for.

Aside, however, from this particular phase of the subject, I think that, simply as an educational principle, specialization of the proper sort, i.e. the utmost possible thoroughness in some one subject or related group of subjects, and the use of this as a centre for the grouping of others, cannot begin too early. From the individual's standpoint, there is in such an education certainly the greatest happiness, and, through that, the greatest profit. It makes him also at the same time a more valuable member of the community to which he belongs. The minds of children come to the teacher somewhat as the blocks come from the quarry to the builder, of diverse sizes, shapes, and textures. The skilled workman does not first reduce them all to one kind, but he takes account of the differences of their natures, and makes of this but a common building stone, of that a column, and of another the capstone of an arch. So the effort to train students all alike through their school course, leaving the cultivation of their particular talents to the college, is not logical and not economical.

In summary, the opinions as to the place of the Sciences in Education that are here advanced are these, — that the Sciences are intrinsically of as great educational value as other subjects, for some minds more valuable and for some less, and hence should be admitted to full equality with them, being required where they are required and they being elective where the Sciences are elective; that since the school curricula are full, the Sciences should be introduced as limited options and the older subjects put upon the same basis; that the election should be on a group system to secure all needful breadth; that schools which cannot afford to teach several Sciences shall teach but one or two, making these the equivalents of elementary courses in the colleges; that specialization should be made the centre of every individual's education. Whether or not the reader agrees with these conclusions matters less than that he shall have definite opinions upon these subjects and actively forward them.

The precise educational value of the Sciences will be considered in the next chapter. Botany is but one of several of them, and they differ less from one another than they as a whole differ from other subjects. From our present point of view they are, Physics and Chemistry, largely experimental in their nature, Zoölogy and Botany, largely observational, and Geology and Physical Geography, requiring macroscopic observation and generalization. Astronomy and Human Physiology are of much less importance, since practically it is difficult to bring students into actual contact with their phenomena; and Experimental Psychology is excluded because of its unorganized state at present, and its recondite nature. Since the three first-mentioned groups thus differ somewhat in the training they give, it will be best for the school to offer at least one from each group. If but two can be had, they should be from the first two groups; if but one, then it may best be from the first, preferably Physics. Those of the first group are most expensive for equipment, those of the second, next, and of the third, least. As to the second group, for training it matters not in the least whether one studies the fixed and food-making Plants or the locomotive and food-destroying Animals, for these differences are insignificant as compared with the resemblances between them as living organisms. Zoölogy has some advantages; the structure of Animals is far more sharply differentiated than of Plants, and throws great light upon the structure

of Man, and hence gives the best basis for the understanding of many of the facts of human physiology,
hygiene, etc.   On the other hand, experimental physiology is easily possible with Plants, many of whose
most important processes are identical with those of
Animals; and Plants are easier than Animals to obtain and keep.   Botany perhaps comes somewhat
closer to our daily interests than Zoölogy, and æs-
thetically it certainly far exceeds the latter.   But which
of the two is chosen in any case may depend upon the
predilections of the teacher or other extrinsic causes.
Often a combination course in Biology made up of parts
of each is offered, which certainly has advantages,
though I am of opinion that, upon the whole, there is
more profit in a full course in one or the other than in a
half course of each.   Most people undoubtedly consider
Botany a much easier subject than Zoölogy, but that is
due entirely to the fact that hitherto only its more
superficial aspects have usually been selected for study.
The very ease of collection, preservation, and identification of Plants has led teachers to magnify that phase
to the exclusion of others, a fact which abundantly
explains the rather low estimation in which as a study
Botany is popularly held.   Where two or more Sciences are studied in succession, practically the most
profitable sequence is as follows, — Chemistry, Physics,
Botany, Zoölogy.

II. WHAT BOTANY IS OF MOST WORTH?

THIS question is here asked solely with reference to teaching, especially in an elementary course. As for the Science itself, and its investigation, no one part is of more importance than any other, for all are of the utmost value, and the field is boundless in every direction. But in teaching, selection is imperative, and it is necessary to find out what will give the best returns for the time and energy expended. Of course this best, or optimum, must always be a resultant between the practical limitations of available time, equipment, etc., on the one hand, and the opinion of the teacher as to what constitutes the best training and most useful knowledge on the other. Laboratory equipment and related matters are treated in other parts of this book; I shall here try to examine what data there may be for a judgment upon the relative educational value of the different phases of botanical study.

A comparison of the elementary courses in Botany offered by the different colleges and high schools, or of the several books of recent date written as guides for elementary courses, shows the greatest diversity of plan. The majority of the courses, particularly in

schools of small equipment, give precedence to the
external anatomy, terminology, and classification of
Flowering Plants, in this following Gray's "Lessons."
Others make much of minute anatomy. Some treat
the morphology and ecology of the higher Plants
with little reference to the lower. Yet others give
special emphasis to the lower groups, making of first
importance a study of these and their relationships.
Finally, a few attempt to combine the most important
parts of each of the others, naturally adding some
experimental physiology. And there are all grada-
tions and combinations of these diverse plans.

Without any doubt each of these courses has advantages, and it is very certain that any one of them
in the hands of an enthusiastic advocate is better than
any other taught with indifference. But it is impossible that plans so diverse can be in the abstract
equally good, and the very fact that each has merits
lacking in the others, shows that each (except perhaps the latest-mentioned) is in something deficient.
Surely, aside from individual opinions, and granting
fair facilities, there must be a plan objectively more
excellent than any we have yet found which shall
utilize the best which exists in the different ones, and
be more complete than any of them. If such a course
could be agreed upon, it would form a most valuable
standard of comparison and ideal toward which to
work. But in addition to these theoretical considera-

tions pointing to a possible optimum, there is another
of great practical importance, namely, since the preparatory schools prepare students for many different
colleges, and since the colleges draw from many
schools, there will be much confusion and waste
unless preparation and requirements fit one another,
which can only come about by at least an approximately standard elementary course. This point is
strongly emphasized by the present efforts being made
by both colleges and schools to secure such uniformity
for other subjects. This approach to uniformity can
come about only slowly, and after much trial and discussion, but it will be much forwarded if every teacher
in publishing his plan would add his reasons for adopt-
ing it in preference to others. No doubt in practice
the subject will work itself out in the form of a series
of standard exercises in anatomy, physiology, etc.,
much as it has already done in some preparatory
courses in Physics, results being judged, not from
examinations, but from laboratory records and note-
books.

The greatest obstacle to the attainment of this more
uniform and optimum plan is, of course, the difference
of opinion amongst authorities as to what should constitute it. Where opinions are convictions founded on
the study of evidence, only full discussion is necessary
to bring about agreement. Unhappily, however, many
of our supposed opinions are but predilections based

on some unconscious prejudice given us by early surroundings or education, and are no more matters of
conviction than is the language we speak or the
country we are loyal to. By constant meditation
upon the excellences of those phases of the science
which he likes best, usually those in which he has
been best educated, one becomes impressed by their
great value for training and as knowledge; and in the
absence of constant comparison with other phases,
one's own naturally comes to seem most important
of all. This view once established, all new facts sus-
tain it; for to him who has put on colored glasses,
all things look of that color. It is, then, particularly
important to endeavor in this discussion to put aside
prejudices based upon the nature of our own education, and to attempt to rise to a standpoint high enough
to give a view over the entire subject. Happily, this
is becoming easier with each succeeding generation
of teachers; for our best colleges are now giving a
thorough and well-rounded botanical education.

All teachers must agree that the optimum course
will be that which combines the best training with
the most useful knowledge. Our inquiry, then, re-
solves itself into these two questions: What phases
of Botany best develop the scientific instincts? and,
What knowledge of plants is educationally most useful to the average man?

Of all scientific instincts, the very foremost in im-

portance is that for exact observation. No others
can be of much value if it be lacking; hence training
in it should be the first care of any course. For
training in observation, anatomy, dealing with actual
structure, is the best possible discipline. In order to
secure concentration upon it, and not to distract the
attention by too many novelties, the earlier laboratory
exercises of any course should be upon objects already
somewhat familiar, with clearly defined characters, and
large enough to need no tools, but only the naked
eye and hand. Answering to these demands, there is
nothing known to me better than large seeds, which
have the further advantages of being easy to obtain
and in condition for study at all seasons, as well as
a logical point of beginning for the study of the cycle
of plant-life. The correct sizes and shapes of these
seeds, the exact kinds and relative positions of all of
the markings on the coats and their relations to the
parts of the embryo inside, the number of the coats,
the full number of parts in the embryo, and the exact
way they are put together, all afford under the skilled
teacher fine materials for practice in observation, a
failure to succeed in which cannot be laid to inability
to use instruments or ignorance of how to begin work.
It is active seeing, not passive looking, which constitutes observation. Later the seeds may be germinated,
and the exact place and mode of appearance of new
structures, the position of newer leaves relatively to the

D

older, how the veins branch and end in the leaf, where the flowers appear, etc., will afford extremely good materials for training in observation, and of the most direct sort. Tools may gradually be introduced as the need for them is felt, at first but a knife for simple dissection, then a lens to help the eye, and later the dissecting microscope to aid hand and eye together. It is only after much practice with these simple appliances that the compound microscope should be introduced, and then in such a way as to impress upon students its true function as simply an aid to vision. To begin a course with objects needing the use of the compound microscope, that is, to introduce the use of the most special tool before eye and hand have had some training by themselves, is not only illogical in theory, but, as I and many other teachers know from experience, wasteful in practice. It produces a long and despairing floundering about from which balance and stability are but slowly regained. Moreover, it impresses a wrong ideal of scientific work, implying as it does that there is some sovereign virtue in elaborate instruments, thus tending to elevate these to a rank above their proper grade of mere aids to eye and hand. After the use of this instrument has been learned, however, microscopical anatomy is one of the best of disciplines for training in observation.

Next among the scientific instincts I would place that for critical comparison and generalization, the

morphological instinct.   It consists both in a power
to compare a series and eliminate what is individual
and unimportant from what is common to a number
and important, and also in a power, by comparison
of different stages of development, to trace back differing forms to their common origin, or similar forms
to their different origins, as the case may be.   For
training in this power, so important in all phases of
human activity, nothing is better than morphology,
the introduction to which is best made through forms
which are large and plain enough to need no tools,
but only the unaided eye and thought.   For this the
embryos in seeds, which show homologous parts under
the greatest diversity of size and form, are particularly
good, especially since they may so easily be traced
through stages of germination and growth where
actual proof of their analogies and some of their
homologies may be found.   Best of all for morphological training, and most used, are the shoot (leaf
and stem), root, flower, and fruit, of the higher plants.
It is true that minute plants also offer extremely good
materials for morphology and anatomy, but they require the use of unfamiliar tools and methods, and it
is better at first to use materials in which the attention need not, by purely mechanical conditions, be
distracted from the real problems.

Next among scientific instincts comes faith in cau-
sality, involving the belief that every phenomenon is

yoked with preceding factors, combined with a desire
to learn what these are.   For the cultivation of this
instinct of causation, anatomy and morphology should,
from the first, be viewed in the light of the factors
determining them, that is, they should be approached
through physiology and ecology.   It is an important
and valuable discipline to study the exact way in
which leaves are built, and to learn their diverse
forms and special modifications; but only a fraction
of the value of this study is realized unless it is made
in the spirit of constant inquiry, which asks why they
are flat, and horizontal, and thin, and greener on the
upper than on the lower surfaces, and why they have
become compounded or altered to spines or provided
with stipules.   It is not, however, enough to simply
ask these questions; the habit of actively seeking the
answers to them must be inculcated.   And the answers
come in part through observation, but more through
experiment. Indeed, so important a factor is experiment in elucidating causation that one may almost
speak of the causative instinct and the experimental
instinct as synonymous.   The cultivation of the habit
of testing the connection of causes and effects by
experiment is therefore a most important part of
botanical training.   An experiment is a definite question asked of nature, and properly follows after all
possible observation and reflection, and is most often
a testing of possible hypotheses suggested by this

reflection. For cultivation of the experimental instinct of this definite kind, there is nothing in Botany to equal simple physiological experiment upon such topics as respiration, photosynthesis, absorption, geotropism, where the object of the experiment is perfectly distinct, and the results obtained are positive and logically conclusive.

Practically, physiological experiments most profitably come along with the particular structures which they best explain. Thus, experiment upon absorption of liquids should accompany study of the structure of the root, photosynthesis that of the leaf, respiration that of growing seeds, etc. In the same way topics of ecology should accompany the study of the structures best explained by them; seed locomotion accompanies the study of seed-structure, locomotion of pollen that of the flower, etc. It is sometimes maintained that general physiology and ecology, particularly the former, should be taken up before anything else, and, theoretically, this view appears correct. But for most teachers this plan would have in practice great difficulties, since the student would not only be plunged at once into a sea of unfamiliar phenomena, but also would have his attention distracted by the use of many unfamiliar instruments and methods. He would finally gain a correct proportioning of the subject, but only after a great loss of time and energy. And, moreover, I think all the value of making physiology thus

form the introduction to structure is realized in studying them side by side, the physiological or ecological
observation or experiment accompanying the observational study of structure. Structure, which is something real and placeable in concrete form before the
student, and already more or less familiar, will probably
always constitute the best skeleton for an elementary
course.

Another scientific faculty of the greatest importance
is that of estimating evidence to the formation of conclusions that are to be held as logically proven, as
probable, as possible, etc. For this, again, physiological experiment is particularly good. Another power
is that phase of imagination known as visualization,
the ability to project before the mind vivid images
of real structures, and to build up complete images
from isolated data. For this, microscopic anatomy is
particularly well adapted; the microscope shows at
any one time hardly more than a single plane of an
object, and it is only from a number of views that
an image of the entire object can be constructed. A
power of visualization is a great aid to generalization.
All of these faculties and powers are elements in
Induction, to cultivate the habit or instinct for which
is, of course, the first object of the science teacher.

Another very important power which scientific study,
with its basis of exact data, is peculiarly adapted to
promote is that for terse, logical, complete expression,

a subject so important that it is treated fully in a later chapter. It is in this connection that descriptive work in Botany, including the use of exact terminology, is held, and I think rightly, to be of great value. On this ground many teachers make much use of blank forms for description, lists of descriptive terms, etc. Unhappily, however, instead of being kept in its proper subordinate position, this kind of discipline is too often given the leading place, and this is at present the greatest defect in botanical teaching in this country. It places instruction in description of a particular phase of plant-structure before the study of the leading facts of plant-life. It is not that this work is not valuable, but simply that it is not the most valuable way in which the student's time and labor may be spent; nor is it at all representative of the present state of botanical science.

Among other advantages inculcated by scientific, and hence botanical, study may be mentioned : intellectual honesty, without which no real scientific work can be done, involving impatience of affectations and genuineness of character; the habit of objectivity and elimination of anthropomorphism, essential to a true understanding of humanity as well as of the remainder of nature; intellectual independence needful to all higher mental work.

If these conclusions as to the worth of the different phases of Botany in the training of the scientific

faculties are correct, it follows that no good botanical
course aiming at real scientific training can afford to
confine itself to any one phase of the subject, but
must treat it synthetically to realize its best possibilities. It is true no ordinary course can carry out this
plan to its logical extreme, but it sets an ideal toward
which to work.

We must next ask what Botany is of most worth as
knowledge to the average man of education. An elementary course must take careful account of this, since
the great majority of students go no further in the subject, and the course must be made complete in itself for
them, as well as a foundation for those who do continue
into higher work. The most important knowledge, I
should say, is that which, when a man looks upon the
world of plants, enables him to know those facts and
principles about them which are most fundamental,
wide-reaching, and illuminating. Such are, — the real
position of plants in nature, how they live, why they
have their shapes and colors, how they are made, how
each one fits so exactly its individual surroundings, and
what are the principal kinds of them. To understand
their real place in nature, one must know at first hand
the process of food-making in sunlight through chlorophyll, and what it means to other organisms; to know
how they live, one must know their leading physiological processes, absorption, respiration, reproduction; to
know how they are made, one must know the nature

of cells and tissues, and the influences controlling their distribution, and also the morphological composition of plants and the possibilities of great variation upon each structure after it is once formed ; to know the meaning of their shapes, colors, and sizes, one must know the leading principles of adaptation to the external world ; to know how each fits its individual situation, one must know the nature of irritability; to know the principal kinds, one must study them by groups, and their relations in a system of classification.

The most fundamental and illuminating knowledge about plants must surely be that which concerns their life ; yet some of its most important phases are those least studied.   Irritability, for example, which answers in plants to sensation in animals, is, next to photosynthesis, their most important power, explaining as it does how the plant directs its growing parts from the very moment it bursts from the seed, how it places its roots, stems, branches, leaves, flowers, and guides all its movements into definite and advantageous directions.   I often reflect with astonishment upon the almost utter neglect of this most important phase of plant-life in botanical education, and this in the face of the fact that its experimental study is comparatively easy.   Yet many other physiological phenomena of well-nigh equal importance are likewise neglected, and even photosynthesis is often ignored.   The lack of attention to ecology is not unnatural in view of its comparative newness, but rich material for its use is rapidly accumulating, and will form a large element in the good courses of the future.

Very important, too, as knowledge is an acquaintance with the different kinds of plants, their modes of life and relationships. This kind of study may well be termed Natural History, which is much broader than structure and classification, for it takes account of habits that explain structure. One limited phase of this subject — the structure and classification of the flowering plants — is more studied in this country than any other phase of Botany, and it is usually supplemented by copious memorizing of terminology, and often by much labor in the preparation of collections, all of which, while not without value, is yet sadly insufficient and unrepresentative of the science. The study of the Natural History of Plants, in order to be valuable, should be comprehensive and cover all of the leading kinds; but since time will not permit this to be done in any great detail, it is needful to select from each group a few of the most typical and important forms, those which best illustrate the relations of the group to others and its position in the plant-world, and to study these carefully. I think it is much more profitable to study *all* of the groups, Algæ, Fungi, Lichens, Bryophytes, etc., in this way than to study more minutely any one or even several of them. By some teachers this study of groups is made of first importance and takes precedence of the study of principles which is here recommended; this is without ques-

tion a very profitable method (though in my opinion not
the most profitable), particularly if it is well proportioned
and covers all groups. Some, however, rather concen-
trate attention upon the lower groups and give a thorough course in Algæ and Fungi. From the point of view
of botanical knowledge for the average non-specialist, I
think this is unwise; for while it is most desirable to
know, for example, Fungi as a group, what their
mode of nutrition is, and reproduction, and their general
relations with the Algæ, and also to know forms of such
economic importance as the Bacteria, or of such prominence as Moulds or Mushrooms, it is not so important
to spend time upon studying the minor groups whose
differences are so slight as to need a microscope to make
them even visible. Knowledge of things he can see with
his own eyes in the world about him is more important
to the student than knowledge of things which he can
only detect by use of special instruments, though this is
true only in general and with many exceptions; but it
emphasizes the value of a comprehensive study of all of
the groups rather than a more special study of a few,
particularly if the latter be of the lower and smaller
plants.

As to the use of Manuals and the determination of
species of the higher plants by their use, — a subject
which forms the major part of the Botany taught in our
elementary courses, — I doubt if this should necessarily
form a part of any regular course in high school or

college.   Its proper place is in the lower schools, where
children learn names easily and with pleasure.   It is safe
to say that ninety-five hundredths of those who are later
taught to use a Manual and make a collection in school
or college never again look at either, and very soon forget how to use the one and where they have put the other.
Therefore, I do not think it is right to take time in which
all students may be taught illuminating and fundamental
facts and principles they will be slow to forget, and in
which they may be gaining a training which will be of
use to them in any occupation of life, for learning how
to look up the names of plants in a book; especially
as any person of fair intelligence who cares at all for
it can pick up the names of common plants in other
ways much more easily.   Some knowledge of the main
divisions of the flowering plants, and of some of the
leading families, comes naturally in the study of Natural
History of the Group of Spermatophytes.   Still, the
ability to use Manuals, and some knowledge of classification of the higher plants, is invaluable to certain
students, especially to those who continue their studies,
and to many to whom it is an agreeable hobby.   Opportunity for systematic study should therefore be given
to such students, but this can readily be done through
voluntary classes of those interested working at odd
times — a system I have myself thoroughly tested, and
use to my entire satisfaction.

Any course embodying the ideas here set forth must

necessarily represent a compromise between many different and often conflicting considerations. Such a compromise, following the recommendations in this chapter, and in every detail worked out in the laboratory, is given in the Outlines contained in Part II of this book. They have two Divisions. The First includes the Principles of the Science of Botany, worked out by following the higher plant through its cycle of seed, seedling, adult, flower, fruit, to seed again, the unfolding of each successive organ being made the basis for the study of the physiological or ecological principles controlling its development. The Second Division includes the Natural History of the Groups of Plants from Algæ to Spermatophytes; here the principles learned in Division I are applied to the understanding of the place in nature of the leading groups and their most important representatives.

The conclusion of this chapter then is this, that from all points of view the most valuable elementary course in Botany is a synthetic one, which confines itself to no one phase or division of the science, but takes from each what it has of most value to offer.

III. ON THINGS ESSENTIAL TO GOOD BOTANICAL TEACHING

THE true teacher of Botany, as of any other subject, is born not made. But a chief birthmark is a determination for incessant improvement.

Certainly the first acquirable quality of a good teacher is a thorough botanical education. This can best be obtained by a full course in some one of our leading colleges which possess properly organized departments of Botany, which, unhappily, some of them do not. In this, as in other matters, it is well to remember that it pays to have the best. Without doubt in the future the teachers of Botany in the high schools as well as the colleges will be thus trained, for there is a strong and healthful tendency for the schools to employ specialists, and also for the colleges to train them even as undergraduates. It is true that in most schools a single subject cannot have a teacher to itself, but it can be one of a very few related subjects.

Next in value comes attendance at summer schools, which several of the principal universities maintain in the holidays for those who cannot attend the winter sessions. The obvious objection to these, that they impose hard work at a time when the teacher should be resting, is not so great as it seems. The change in occupation, surroundings, and companions brings so much relief in itself, that the work is less felt; and besides, if the learner's spirit is of the right sort, and the teaching is of the true quality, the pleasure of it all should go far to lighten the labor. In my own experience, too, I have found that there is more rest in change of occupation than in absence of it. Perhaps the mind is in this like the soil, that it does not need to lie fallow, but can continue to bear without exhaustion if given a wise rotation of crops. Of all the excellent summer schools in this country, however, that which is in my opinion the most profitable to the teacher is the Marine Biological Laboratory at Wood's Holl, Massachusetts, which belongs to no single institution, but to them all. Here, in addition to the excellent courses and the great facilities, there are opportunities unrivalled in this country for that acquaintance with co-workers and specialists, and that scientific atmosphere, which are worth to the teacher as much as the instruction itself.

After summer schools, and long after, comes study by one's self. But this to be efficient should be actual practical laboratory study done under the advice and criticism of some specialist. Study without guidance is sure to be full of gaps and bad in its proportions. Correspondence courses, which some institutions offer, are better than nothing, but are a very poor substitute indeed, especially in the sciences, for contact with a skilled teacher. No doubt books could, and in time will, be prepared as guides for those who must study alone, but at present hardly any exist.[1] Much good may be obtained from reading, but only as supplementary to the actual study of botanical objects, never as a substitute for it. To attempt to truly know Botany from books is like expecting to acquire the advantages of a European tour from the reading of guide-books, or like trying to form intimate friendships through the exchange of letters. In another part of this work (VII. On Books) will be found further suggestions upon reading, and lists of books that can be recommended.

Whenever advice is needed about books or any other botanical matters, the teacher should not hesitate to write to some botanical specialist, as, for example, the Professor of Botany at the nearest large university. Most specialists take pleasure in assisting any earnest inquirer, and many of them welcome this method of extending their own usefulness as teachers.

There is one feature of the education of the teaching botanist so important as to deserve particular emphasis, i.e. the performance of some work in original inves-

---

[1] Strasburger's " Das kleine botanische Practicum " (English translation by Hillhouse, under the name of " Practical Botany ") could thus be used, and with great profit, in plant anatomy ; and Bailey's " Lessons with Plants " would form a good guide for such study in general elementary work.

tigation.   Good scientific teaching consists above all
things in the application of that natural, independent,
inductive spirit which is the only basis of scientific
progress; and only he can properly apply it who has
himself experienced it.   The college graduate, if his
undergraduate work has been of the right sort, has
had some of this training, for undergraduate courses
at their best are carried on in the investigation spirit;
they constitutute a series of problems which subjectively are original investigation, whence the transition
to objective investigation in the graduate courses of
the university is easy and natural.   But in addition to
this, a year or more of graduate investigation will tell
very greatly for the better quality of the teacher's
work, especially if he continues investigation afterward.   Where graduate study is impossible, the teacher
will do well to take up some problem for himself.
Under present conditions it is becoming nearly impossible to do good original work in anatomy, morphology, or physiology away from the facilities of the
universities; but happily there is a great field with
abundant problems in the comparatively new study of
ecology.   The habits of even the commonest plants,
especially in their relations to the other organisms
about them, are very imperfectly known; and there
is not a section of any country in which there is not
inviting opportunity of this kind.   Especially attractive
at the present time are the problems of ecological

E

plant geography,—the study of the reasons why each
plant stands where it is, and is of the form, size, and
texture it is.   Very attractive, too, are the problems
in modes of locomotion of our common plants, and
in the mechanisms of cross-pollination of many of
them.   The construction of a local flora in which each
plant is not only listed, but located ecologically, is
everywhere possible, and would be both scientifically
and subjectively profitable.

There is yet another line of original research open
to the teacher, — the investigation into better and more
economical ways of utilizing the science in education.
There is here opportunity for doing very great service in the selection and demonstration of the most
profitable topics, in the invention of simpler and more
logically conclusive experiments for proving the most
fundamental principles, in the discovery of more illustrative materials for the different phases of the study,
in a word, for the deduction of such outlines as are
offered in this work, and for their great improvement.
And this line of investigation is as legitimate, as difficult, and as important for the advancement of botanical
science as is the elucidation of vegetative points, chro-
mosome numbers, or transpiration currents.

Not only does the study of original problems increase
the teacher's power, but it adds intensely to the interest
of his life and profession.   The perennial freshness
which accompanies constant progress goes far to coun-

terbalance that monotony of yearly repetition of class-work which is the greatest drawback to the life of a teacher.

A thorough botanical education stands so far above all other needs for good botanical teaching that any consideration of the cultivation of special qualities, or of the use of special methods, seems hardly to belong in the same chapter.   There are, however, qualities which may be cultivated to the great profit of the botanical teacher, and methods which, like other fine labor-saving tools, enable the skilled workman to do yet better work, and these it will be worth while briefly to consider.

Many of the qualities essential to good botanical teaching are, of course, the same as are needed for success in any teaching; these are the qualities constituting the teaching temperament.   This consists in a deep-seated pleasure in the exercise of guiding minds from ignorance to knowledge, and in seeing the light dawn through darkness; in a power of positive self-reliant leadership; in ability to project one's self into the student's mental position; and in a personality that can win respect and affection.   Of all these characteristics, sympathy is one of the most important; for the good teacher is, first of all, a mental physician of the truest sort, diagnosing each individual case, and fitting its proper treatment to it.   He is a leader, and not a driver.   He is always an uncompromising though genial critic, using sarcasm only for otherwise incorrigible cases. He diplomatically makes use of all devices for arousing interest and holding attention. Especially is he ever investigating, experimenting, and improving in his teaching, reading newer books upon it, and keeping in touch with educational progress as shown in the educational journals. It is, indeed, only by constant advance that he can escape that mental drying-up, which is the greatest danger, and too often the most obvious badge, of the teaching profession. And he has a deep respect for his profession, views it as his life work, and upon every possible occasion champions its interests.

There are one or two practices in general teaching which I think are not only bad in themselves, but particularly so in scientific teaching. Of these, one of the worst is the devotion of a larger amount of time and energy to the dull students than to the brighter ones, a practice rendered almost inevitable by the systems of grading now widely in use. It is always to be remembered that education cannot undertake to make a new man, but only to make the best of what man there is, and the good student is as entitled as is the poor one to have the very best made of him. Children are born most unequal in mental powers, and to attempt to bring dull pupils up to near the standard of the bright ones is not only a sad waste of energy upon an impossibility, but is as well a great injustice to those who are brighter, for they are de-

frauded of their rightful share of the time and energy
of the teacher. School and college are, after all, but
a preparation for the world outside, and the principles
controlling human society generally should surely be
used as a guide in education. The world at large
does not leave its brighter members to shift for themselves and devote itself to the elevation of the dull
ones, and it will save the dull students and their families much disappointment later if they are allowed to
find their own level in school or college. I do not
mean that the dull students are to be neglected by
the teacher, but simply that they are to receive only
their fair share of attention, and that it is just as
much the teacher's duty to take time to lead on the
best pupils into still higher achievement as to urge
the duller to greater efforts.

Again, it cannot be too often nor strongly emphasized in education, as in development of the body,
that it is through effort that strength is gained, and
through responsibility that character is formed. The
great refinement of methods in recent years has had
a tendency more and more to shift the responsibility for learning from the student to the teacher,
and to make the student consider that his duty ends
with a blind obedience to the teacher's wishes. This
is a very wrong attitude; the responsibility for learning should be kept upon the student, who, therefore,
should not receive aid and admonition at every step

in his work, but be obliged to complete certain portions, or topics, to be judged en masse, and which he must therefore plan to do for himself. The teacher should try to cultivate the idea that it is a teacher's duty simply to provide the opportunity to learn, while the responsibility of taking advantage of this opportunity rests upon the student. Of course, this applies rather to the better students; it will always be necessary to force the poorer. Again, information and training are still too often confused, and the first place too often given in teaching to the former, and this despite the fact that "what a man can do is more important than what he knows" has become a commonplace of education. Training is immensely more important than information in education, for many reasons, and amongst others for this, that the acquisition of information is a power that follows as a matter of course upon training, while there is comparatively little training in the acquisition of information. The trained mind naturally, and without effort, assimilates information and transmutes it into knowledge, while the untrained simply stores it en masse to the limit of its capacity, all kinds together. Information, too, which is simply, as it were, laid upon the mind, may soon be forgotten, while training, consisting in a change in the mind itself, always remains. The aim of every teacher should be, training before information, or, in the words of President Eliot, every teacher should *Train for Power*.

There are some principles of scientific education not yet everywhere understood as they should be, and the first of these is the absolute necessity for laboratory instruction and the almost worthlessness of book work alone. To try to realize the value of scientific study from books without laboratory work would be, as I have before said, like trying to derive the advantages of a European tour from the reading of guide-books, or the effort to form close friendships through correspondence, comparisons which the teacher may well use in answer to those who doubt the necessity for laboratory work.

Again, it is often maintained that a chief object of scientific study is to increase a love of nature, or to produce a greater reverence for the works of the Creator. On the contrary, its objects are utterly different. If these things follow incidentally, so much the better, but they are not a leading object. All scientific teaching should be, first of all, as clearly cut, distinct, logical, statistical as possible, and anything permitting haziness of ideas should be rigidly excluded. Its object is to help to train the intellect to be, as Huxley puts it, "a clear, cold, logic engine, with all its parts of equal strength, and in smooth working order." For these reasons I think that both poetry and religion should be kept out of scientific laboratories. In the first place a large proportion of the poetry introduced by teachers and taken up by pupils is false in sentiment and of a

weak and washy nature, and in the next place both
teachers and students use it as a cloak for hazy ideas
and a lazy release from difficult problems.   It has been
my observation that those teachers who talk most of
the wonderful works of nature, and of loving it for
the Creator's sake, and who put verses from the leading
poets upon the laboratory blackboards, are the weakest
in the scientific quality of their teaching. A great
deal of nature study in the schools is also blighted by
this weak sentimentalism.[1]   The argument that poetry
should be used in the laboratory to stimulate the imagination, since the imagination is of great scientific use,
is utterly fallacious, for the kind of imagination used in
science is visualization and generalization, which are
injured more than they are helped by the lighter plays
of fancy, the metaphors, and the impressionism of real
poetry.   Mathematics is a much better training for the
scientific imagination than is poetry.   Similar objections
apply to religious ideas, which in their use by the average teacher are liable to be misinterpreted, and of more
ultimate injury than good.   When I say that poetry and
religion should be kept out of laboratories, I by no means
say they are to be kept out of education; on the contrary, I think they should be carefully inculcated, but,
like the sciences, they should be studied from their own

---

[1] A needed warning on this subject is given in "Sentimentality in
Science Teaching," by E. Thorndike, in the *Educational Review* for
January, 1899.

materials under those trained in them.   There is inherently no more reason why they should be dragged into scientific laboratories than into mathematics or athletics.

We shall now consider, more specifically, good procedure in the botanical laboratory itself.   The arrangement and equipment of the laboratory room and the use of outlines, note-books, etc., are considered in later parts of this book.   The number of students in one laboratory division should not, at the most, exceed twenty-five or thirty, unless an assistant is available, when it may be larger.   Two-hour periods are the best for beginners.   Students do not become weary in that time, and shorter periods are uneconomical on account of the time lost in getting the work started and things put away at its completion.   The amount of work laid out for each period may best be adjusted to rather above the average student;   and more exact and detailed work may be expected from the best members, while the poorer must be permitted to do it much less completely and perfectly.   In each new laboratory period all students should start the new topics together, uncompleted work of earlier periods being made up in time outside of regular hours, for which, as well as for extra voluntary work, the laboratory should always be open.   The actual laboratory work is best managed on the *practicum* plan, that is, the students are all working upon the same problems or approximately so, and the teacher goes about among them, giving individual en-

couragement or criticism, and from time to time, as
the progress of the work requires, making suggestions,
explanations, or summaries to the class as a whole, and
closing each period by a summary of the work of the
day. This plan does not in the least interfere with
the independence and value of individual work by the
students, and, on the average, seems to me the most
economical for conducting elementary classes.

In the matter of order, etc., in the laboratory, the
teacher must be careful to preserve the free and home-
like spirit so essential to natural methods of working.
Considerable freedom of movement and conversation
must be allowed. Indeed, a silent laboratory would be a
most depressing place. Students should, of course, be
expected to keep their own places and instruments in
good order, and to take a corporate pride in the appearance of the laboratory as a whole. They should learn
to put away every tool after using, as an integral part
of the very act of using. They should learn also to
work in physical comfort and with deliberation, and
to be exact and neat in all their work, doing it not
simply well, but the best they can. But of course
order and neatness can be carried too far ; and there is,
as in other things, a certain *optimum* of neatness and
order about a laboratory that should be aimed at,
rather than the *maximum*, which demands an undue
and uneconomical expenditure of labor.

In the laboratory work everything possible should be

done in the independent investigation spirit. The student should be led on by having each new thing placed before him in the form of a problem, so arranged that its solution comes just within his own powers. In general, nothing should be told a student that he can find out for himself, though with beginners, where everything is new and unfamiliar, this principle must be followed with caution. There are many occasions on which it is best to tell a student minor things outright to help him to the solution of important questions; and there are other occasions when leaving him unaided would result in discouragement followed by a distaste for the subject. The best principle in such cases is to ask a question or give a suggestion in such a way as to allow the student the pleasure of finally solving the difficulty for himself. It is in such points as this that sympathy and judgment count for so much. The teacher will, of course, constantly use such common pedagogic devices as proceeding from the known to the unknown, and of recalling to a student what he already knows as a basis for building new knowledge upon. Another important principle is the refusal of the teacher to examine any piece of work until it is as complete as the student can make it. If the teacher is willing at each step to tell the student whether he is right or not, responsibility is shifted from the student, who will simply do the mechanical work, and let the teacher do the thinking, thus losing that training in self-reliance

which is one of the most valuable features of his education.   It is true the student will in this way make
many mistakes and less apparent progress than on the
other plan ; but in this world there is nothing from
which we learn so much as from our mistakes, and it is
by constant struggling and effort that the mental fibre
is strengthened.   Again, it is important not to supply
information, methods, terms, nor tools until students
have been made to feel a need for them.   Such things
then have a meaning, and make an impression upon the
memory, to an extent impossible when they are supplied without this connection.   Of course all laboratory
work is to be carefully examined after it is completed
by the student, and should be marked when approved.
In my own experience I have found it profitable to
place a small oblique mark at the lower outer corner
of each page when it has been examined, which is
made a cross when the page is finally satisfactory;
and the responsibility of having all their pages completed, examined, and checked is thrown upon the students.   This examination of the work is best made
with the student, not apart from him.

The teacher will always, of course, keep in mind
the main object of the laboratory work, i.e. the cultivation of the scientific inductive habit of mind with
the end to forming a scientific instinct.   In another
chapter (Chapter II) I have tried to trace the characteristics of this scientific spirit.   It consists in the

habit of honest, disinterested observation, in discrimi-
native comparison, and in generalization in proper
logical degrees of truth. Every effort should be
made to cultivate a distrust and dislike for conclusions based upon too scanty data ; a desire to go
always to the original sources of information ; and a
preference for the evidence of one's own senses
above any other source of knowledge. The teacher
may measure his success by the degree of mental
independence he arouses in his pupils. Speculation
should be encouraged, but not allowed to escape the
rigid control of facts. Every possible play should be
given to each student's individuality. The one who
has a taste and a knack for experiment should be
encouraged to become the class authority upon experiment, and another the class artist, and so on.
Whenever possible, they should be allowed to feel
the pleasure and the stimulus of such authority.

In judging the work done by the students, the
teacher should always remember that on the average
the most profitable point for which to aim is not the
maximum, but the *optimum*. It is a fact in education, as in physiological and economic phenomena,
that the return for labor expended increases up to a
certain point, beyond which any further advance is made
at a disproportionately great cost. It is this best, or
optimum, point the teacher should in general seek ; and
he should not compel his students to follow refine-

ments too far. On the other hand, it is true that in one's own specialty it is the maximum that is profitable, and in the face of competition in the world outside it is the maximum that most men are forced to, and this is the logical end in art, music, etc. While, therefore, the teacher should be content with the optimum from the average student, there will be cases of special talent where it is wise to encourage individuals to the attainment of their maximum, their very best possible.

The good teacher, too, will not be above employing many little tricks and devices to arouse interest, keep attention, and encourage application. As the diplomat and the politician play upon the peculiarities of "human nature" and attain success by their knowledge of it, so may the teacher. But he should ever remember that Botany is not an end in itself, but that its highest aim is to contribute to human welfare and happiness.

An important part of good botanical work is experiment on physiology. Practically this is difficult to work into the regular laboratory hours with large classes, and I have found, after trying different plans, that it can best be managed in demonstration hours, when all the students may be present and give their undivided attention to it. The teacher should then set up the experiment before the class, carefully explaining, or rather letting them work out from his remarks, the logic of each step. Each student should then for himself observe and record results, and deduce conclusions as if the experiment were entirely his own. It is particularly necessary that the students understand the exact logic of each step, and that their records should bring it out clearly. Their records, too, should express and keep perfectly distinct (*a*) the object of the experiment, (*b*) the method and apparatus employed, (*c*) the results actually observed, (*d*) conclusions. Results of all experiments should be expressed precisely and quantitatively whenever possible, not only because of the greater scientific value of results of this kind, but also for the sake of the pedagogic value of exact measurement, which is very great.

The close acquaintance the teacher forms with the student in the laboratory makes examinations as tests of knowledge unnecessary, while regular essays may be made to give to some extent that other chief value of examinations, namely, review. Still, examinations for insuring thorough knowledge of the theoretical work are desirable, and quizzes, etc., have their value, varying with the personality of the teacher. Since laboratory work gives a knowledge of but a few types, and since some of the most important topics cannot for practical reasons be studied in the laboratory at all, considerable theoretical, as distinct from practical, work is necessary. This is

in part given through reading, of which I shall speak
elsewhere, and it may well be given in part by lectures or their equivalent. Lectures following the
laboratory work in particular topics, which are then
treated comprehensively and correlated with others
and with the general subject, are certainly most valuable, and full of meaning to students after their
actual practical study, though they are of small value
without it. The lecture itself should be a study in
induction and proportion, as fully illustrated, interesting, and suggestive as possible. I have found in
my own experience that the best balance between
the different kinds of instruction is, — two two-hour
laboratory periods (actually more for most students),
and one lecture and one demonstration (or recitation)
hour a week.

Very important, too, are field excursions, the opportunity for which varies greatly. Theoretically, it might
seem better if most botanical study could be done
out of doors, but practically the greater part of it demands tools and other facilities, including physical
comfort, unobtainable away from a good laboratory.
In the excursions the teacher will of course direct
attention to the larger phenomena of adaptation, the
topography or physiognomy of the vegetation, the
plant associations, etc. This kind of study will become much easier and more profitable in the near
future as the subject becomes more fully systema-

tized, and good books on it become accessible. It is especially important not to allow too great a number of students to go together on these excursions, and in my own experience not over ten can profitably be taken at any one time. The collecting instinct, so invaluable to the naturalist, should at such times receive every possible encouragement, and later will be found suggestions upon its utilization (in Chapter VI).

Finally, it is well for the teacher to teach as far as possible by example, for here, as elsewhere, it is better than precept. It is an inspiration to students to see their teacher himself a student always striving to learn and advance.

F

IV. ON SCIENTIFIC RECORDING, — DRAWING AND DESCRIPTION

In the preceding chapters I have tried to make plain the real aims and profitable procedure in laboratory study. There is one phase of the latter, however, of such importance as to require separate treatment, namely, the making of scientific records.

Exact recording of the results of laboratory work has several values. It is of great utility in general education for the training it gives in preciseness and proportion in exposition of original data. Again, it imposes direction, definiteness, and completeness in observation and reasoning. Finally, and pedagogically most important, it enables the teacher to make sure the student has actually and fully worked out his topics. A clever student may by verbal answers alone convey the impression that he has seen an object fully, when in fact he has seen it but superficially; but he cannot make a scientific drawing or description of an object until he has first seen it accurately and completely, and realized its construction.

The aim of the student in recording the results of his study upon any topic should always be to make his record a piece of good scientific exposition, a model of concise and accurate conveyance of his ideas to another. In this he is to follow the example of the best scientific monographs. For this purpose both drawings and descriptions in words are needed, each expressing something the other cannot bring out so clearly, the two supplementing and not duplicating one another. From the teacher's point of view, however, the drawings are much the more important, since from them he can most readily understand the student's progress. Ability to draw, therefore, is an important element in a student's scientific education. To realize, however, the full value of drawing, it is necessary that this shall consist not in the making of pictures correct in perspective and fine in finish, but in diagrammatic drawings that convey to the mind of the beholder accurate conceptions of the real construction of the object represented. A diagram, even if utterly unrecognizable as a picture of its object, if it correctly represents its structure with the aid of some words of explanation, is a far better scientific drawing than one which arouses admiration by its fidelity to nature as a picture, but fails to express actual structure. If a drawing can be at one and the same time an accurate diagram of the structure of an object, and a picture giving a true impression of its appearance, so much the better; and indeed this is the ideal in scientific drawing. But diagrammatic accuracy is its first quality.

Drawing in the laboratory should be begun only after observation of at least the main features of the object has been completed, though the very act of drawing will call attention to features otherwise overlooked. Drawings should at first not be idealized or pieced out from several specimens, but rather should be accurate delineations of a selected typical specimen, which as soon as possible the student should be taught to select from several presented to him. In the very first lesson, he should be given a fair object, and told to first study and then draw it without help, himself selecting the number of views, etc., necessary to illustrate it fully. Most students under these circumstances answer in despair that they cannot draw. This answer is a sad commentary upon our modern system of education, which so largely neglects this most natural, elemental, and valuable discipline, thus depriving the student of training in an additional and most vivid mode of expression.[1] Of course all students must be required to try to draw, and if at first perspective, shading, etc., are discouraged, and correct outlines alone are insisted upon, all find that they can draw somewhat, and many find an unsuspected power in themselves of drawing well. Certainly, the

[1] An exception should be made in favor of the drawing accompanying nature study in many schools. But this, as manifested in the work one sees in school exhibitions, is much less valuable than it should be, for it leans too much to the impressionist and too little to the diagrammatic side. This is, of course, because the teachers are not trained in scientific drawing.

talents of individuals should receive the greatest
encouragement and stimulation, and if some can
accurately shade so as to make their diagram a good
picture, so much the better. But at first the drawings
must be, above everything, clear accurate diagrams of
the actual structure. To this end every line and spot
in them should represent something in the object, and
no spot nor line allowed to the equivalent of which
in the object the student cannot point. Moreover,
outlines should be complete, and no loose ends, nor
hazy joinings, nor dim angles, should be permitted.
Such imperfections generally correspond to loose, hazy,
or dim ideas, which it is one of the chief uses of the
drawing to remove and to replace by clear and sharp
conceptions. It is for this reason the generalized diagrams, to be spoken of later, are of such great value.
I have found that "rough drawings," sometimes recommended, are of very little use, and the impressionist
kinds, often really beautiful, made under teachers
untrained in scientific methods, are little better. The
true diagrammatic drawing takes but little if any more
time, and is many times more valuable. Indeed, a
mere "drawing" of an object, *i.e.* a representation
of its appearance to the eye, a reproduction of the
impression the object makes upon the beholder, has
very little if any scientific value in connection with
laboratory work, and is not worth the time it takes.
Such drawings are in place in a drawing class, and

even in certain phases of general natural history study; but they reflect not at all the clearly cut ideas which should characterize the activities of the laboratory. Samples of clear diagrammatic drawings may be seen later in this work (Figs. 12, 13, 14). Practically, the best way to make these freehand drawings is first to outline them very faintly in pencil, and then alter this outline until it corresponds to that of the object, after which a single, firm, complete, even line can be run over it, and all the lighter lines erased. All mechanical helps, such as rulers, compasses, etc., should be allowed when they contribute to accuracy.

An important principle in making the drawings to illustrate the structure of an object is economy of number; as many drawings should be made as are necessary fully to illustrate the object, and no more. Thus, for a seed like the bean (Fig. 12), two drawings are sufficient; an end view would bring out little if anything not already in the other two. One view of an object need not duplicate what is already in another, though different views of the same feature should always be included. The extreme aspects of an object should be chosen for representation, *i.e.* a face or edge view should be an exact face or edge, and not a quartering view. Of course, different drawings of the same object should be perfectly consistent as to size, form, etc.

The scale of the drawing in comparison with the

original object is very important, and should always be expressed. This is usually done in good monographs by a fraction ; if the drawing is one-half the size of the original, the fraction $\frac{1}{2}$ should be placed beside the drawing; if the drawing is twice the size of the original object, it is expressed by $\frac{2}{1}$; if the same size, by $\frac{1}{1}$, and so forth. The best general rule as to scale is to make the drawing as small as will allow all features intended to be represented to be clearly seen. If, however, making clear certain of the smallest features would make the entire outline very large, it is better to make two drawings, one showing the details only upon a larger scale. It is well to give the students small pasteboard rulers, preferably on the metric system, which can be kept in pockets in the back of the laboratory books, and used for making the scale of the drawings correct.

The different features of the drawings should be carefully labelled to show their names. The exact spots to which the names apply should be shown by fine ruled dotted lines, as in Figs. 12, 13, 14. In books, for appearance' sake, usually only letters are thus attached to the drawings, and the corresponding names are given in an explanation below or in the text. But in laboratory work I have found that the extra neatness of this plan does not compensate for the loss of time required of the teacher to look up the explanations, and I think it much better to

label the drawings with the names directly, as shown
in Figs. 12, 13, 14, where the whole subject is visible
at one glance. For this labelling a compact vertical
writing, or even printing, is desirable, and should
be cultivated when wanting, and a compact writing
is pleasing, too, for the notes. When one set of words
can be applied to two or more drawings, as in Figs.
12, 13, 14, it is an advantage, but of course is not
essential. Where drawings do not fully explain themselves, they should also be labelled beneath by descriptive words, such as "face view," "transverse
section," etc. Different drawings of the same object,
unless their connection is perfectly obvious, should
be kept in correlation with one another by proper
cross-references. Of course neatness and artistic
effect are desirable qualities in all work, and some
attention may be given to placing the drawings well
on the page, away from the margin, with the long
axis upright, and to leaving plenty of room between
different ones of the same object, and between different topics, etc.

In all of these respects, i.e. completeness and clearness of outline, economy in number, scale, labelling,
neatness, it is pedagogically a very good principle to
let the students at first do the best they can unaided.
After they have done their very best, they are in a position to fully understand and profit by the teacher's hints
as to how they may do still better. Instruction on

these points after their own efforts have made them
feel the difficulties has many times more meaning than
it has before they have themselves tried.   It is important, however, not to confuse them by too many
suggestions at once.   It is much better to point out
improvements in but one or two respects at a time, and
thus come gradually up to a high standard.   The
earlier drawings will on this plan be incomplete, and
they may subsequently be brought up to the higher
grade, or left as a record of progress, not without its
value.   From the first, it is necessary to insist that the
laboratory work shall not be made a drawing lesson.
The laboratory hours are for observation and comparison, and time for outline drawings only can be
taken ; all refinements should be added outside of
these hours.

Drawing with the microscope, after the use of the
instrument is once learned, offers few difficulties, since
the objects are seen in but one plane.   In drawing
tissues, it is a good plan to shade all walls, and leave
intercellular spaces and cavities blank, even in cases
where, as in cross-sections of bast fibres, the reverse
would make a better picture of the object.   Here, also,
diagrammatic clearness is the highest quality of the
drawing.

After the first principles of scientific drawing as here
outlined have been grasped by the student, the teacher
may well give from time to time some instruction upon

the simple use of shading, etc. In this the teacher, as
well as students, will gain great profit by a study of
good models, where shading has been very effectively
used, as in the best figures in their text and reference
books. A particularly splendid model is to be found in
the illustrations to Sargent's "Silva of North America,"
where drawings of seeds, twigs, leaves, etc., such as are
taken up in the laboratory, may be found. Kny's series
of wall diagrams also offer excellent models. It is a good
plan also to have students copy at times into their note-
books good diagrams from the Kny series, or from good
books, especially where an important topic is being
studied with poor material. A drawing copied from
a good source is a better record of an important topic
than no drawing at all, though of course this must be
resorted to but rarely, and then only after assurance of
a perfect understanding of the diagram by the student.

Drawings will ordinarily be made in lead pencil
(Faber IIIIIIII (4II) I have found best), but there are
many advantages in finishing them in India ink. The
drawings may thus be saved from rubbing through hand-
ling of the books, are more permanent, clearer, and of
better appearance generally. Liquid India ink and fine
mapping pens should be used. Shading can be given
either by fine dots made more numerous for a deeper
shading, or by very fine lines also made more numerous
for a deeper shading, or even given by pencil with an
ink outline. With my own students the use of the ink

is made voluntary, and most of this work must be done outside of the laboratory; but almost invariably the best students, after they have once tried it, take to its use altogether. The improvement made by use of the ink tends greatly to foster the very desirable pride of students in the appearance of their books or notes. Of course the outlining must first be done in pencil, the marks being erased after the ink has been added. Every encouragement should be given to individual artistic tastes in drawing, even to the point of allowing some use of color. But it is constantly necessary to guard against the eclipse of the naturalist by the artist, and the beautiful drawings must be allowed to be no less accurate than those which are merely diagrammatic. Students should be encouraged to work in physical comfort and with a feeling of leisure, with independence yet readiness to profit by the excellences of their neighbors. It is well to allow the poorer students frequently to see the drawings and notes of the better ones.

For the drawings a good smooth paper, which will take both pencil and ink, is necessary. It should never have a perfectly smooth nor glossy surface, nor yet be rough like that used for sketches by artists. The kind called ledger paper is very good. There are different methods of keeping the drawings, one of the commonest being to make them on separate sheets of drawing cardboard of uniform size (usually 6 × 4 inches) which are

then kept in a simple cover. For advanced students
this does very well, but is less excellent for beginners.
It is very difficult to keep notes and drawings together
in this way, and it allows of easy loss and constant
disarrangement. After trial of many systems I have
concluded that a book is best, and have invented a
special laboratory book which I have used for three
years to my great satisfaction. It is made of the best
quality of ledger paper $8\frac{1}{4} \times 6\frac{3}{4}$ inches, ruled on the
right hand page for notes and unruled on the left hand
for drawings, and is strongly bound in linen. It is made
by the Cambridge Botanical Supply Company, and a
sample is sent by them to teachers. Experience shows
that a thoroughly good book of this kind pays in many
ways, and particularly in the increased care students
give to the neatness and completeness of their work.

There is one kind of drawing of which the value grows
upon me year after year, namely, the generalized morphological diagrams worked out in colors, often called
for in the Outlines in Part II of this book. (See particularly Figs. 15, 28, and their explanations.) To work
them out correctly necessitates the greatest clearness
of ideas, and inculcates comparison and generalization
of the highest value. Indeed, such diagrams demand
thinking of mathematical exactness and clearness. The
coloring to show morphologically identical structures can
be added by water-colors, or by pencils which may be
bought in small boxes containing six colors.

Supplementary to the drawings, and necessary to correlate these, and to bring out features which they do not, are the notes or descriptions. These should be as condensed as possible, both for the effect upon the student's composition and also for the convenience of the teacher who has to examine them. They should not contain anything that can be clearly shown in the drawings. They should usually be complete sentences, and perfect in their English, terse and expressive. Whenever possible they should be thrown into tabular form. Drawings and notes should of course be mutually intelligible and consistent, which is the more easy if abundant cross-references are used. The two are much more effective if kept opposite one another, as they may be in such a book as has been recommended.

Of great importance for review, for generalization, and for securing correctness of proportion are synoptical essays, which should be called for under each topic as soon as it is completed. These essays should be strictly limited in length, yet required to include all phases of the subject of any importance; thus is conciseness and directness cultivated. It would probably be found advantageous to make arrangements whereby these essays could also count as work in English composition. It is not at all intended that the essay shall simply repeat what is already carefully recorded in the laboratory books; it is rather a comprehensive but synoptical outline of the entire subject based upon all sources

of information, — laboratory work, reading, lectures.   It
is primarily a study in proportion and in correlation.
After the students have done their best with their first
essay, it is well for the teacher to read them a selected
one or even one composed by himself ; and to illustrate
this point there is given in Part II, Section 3, one that I
have read to my own students after they have completed
the study of the seed as called for in the first three Sections of the Outlines.   Practically, I have found it most
advantageous to have separate books, uniform with the
laboratory books, for the essays ; the latter are, of
course, corrected and returned to the writers.

V. ON LABORATORIES AND THEIR EQUIPMENT

BOTANICAL laboratories are of many sorts, from those built especially for the purpose by some of the greater universities down to unaltered schoolrooms; but all have this in common, that the room and its furniture are of far less account than the person who directs them. In other words, it is more profitable to give a good teacher to a poor laboratory than a good laboratory to a poor teacher. Laboratories, like methods, are fine tools for skilled workmen, and they give but indifferent results in the hands of those untrained in their use. Proper laboratories every teacher should strive for; but he is not to suppose that good work must be put off until he achieves them.

Many universities, some colleges, and a few high schools now possess good botanical laboratories, and if a teacher has the opportunity to direct the building of a new one, he should visit some of these and ask advice of their directors. He may obtain their addresses by writing to the Professor of Botany in the principal university of his State. But the following points will also be of use:—

First, of course, is the room, in which a prime requisite is abundant light. This implies many and very tall

79

windows, which, if all in one wall, should preferably face
the north in order to avoid exposure to direct sunlight;
but this point is really of no great consequence, since
thick white shades perfectly temper the direct sun, and

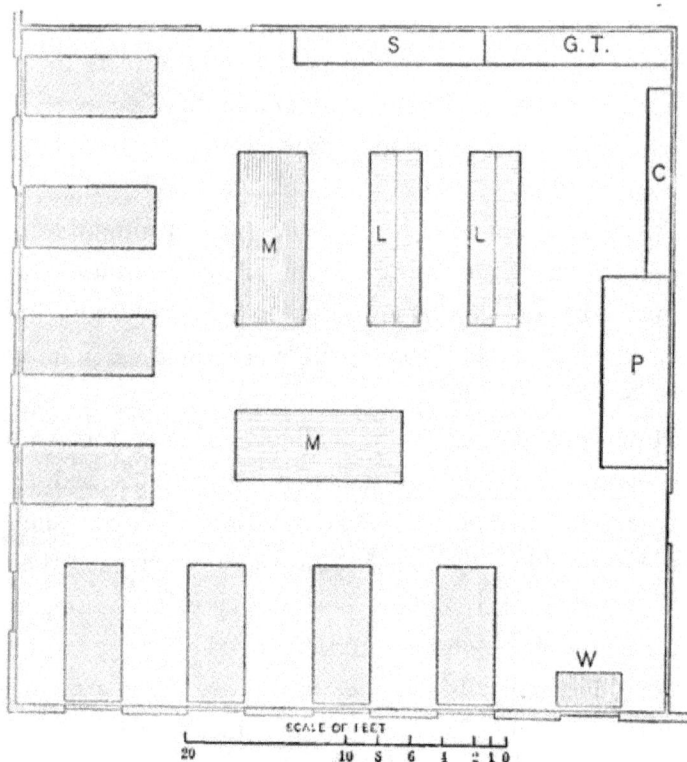

FIG. 1. — Plan for a square laboratory, lighted on two adjacent sides. *C*, case
for museum specimens; *G. T.*, gas and tool table; *L*, lockers for students'
effects, etc.; *M*, tables for materials, etc.; *P*, teacher's platform, with blackboard; *S*, sink; *W*, Wardian case.

in short winter days it is an advantage to have the windows face in the lightest direction. The walls should be

tinted white or nearly so, and the furniture made of light-colored wood. Floors should be solid and dust kept out as far as possible. Unless the windows are very large it is best to use but one row of tables, of which, for elementary work, the most efficient and economical distribution known to me is that shown in the accompanying figures (Figs. 1, 2). Each table stands opposite a

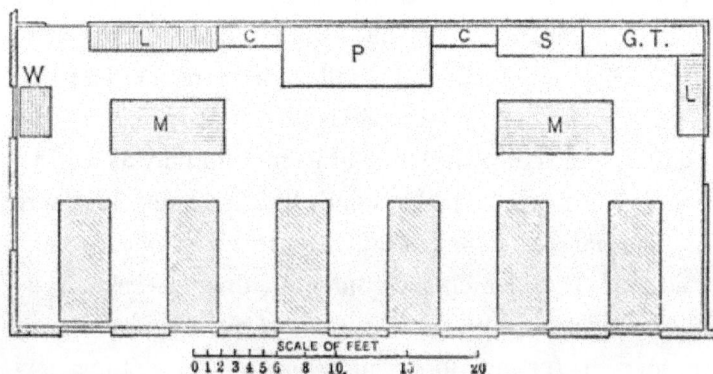

FIG. 2.— Plan for an oblong laboratory, lighted on a long and a short side. Lettering as in Fig. 1.

window, and is used by five students (or three if room allows), two on a side and one at the end. Rooms are often of square shape and lighted from two adjacent sides, in which case the arrangement shown in Fig. 1 is good. For a long room the most economical arrangement is that shown in Fig. 2. Where more than one row of tables must be used, it is best to place them in a second row, each exactly in line with one in the first and four feet from it. The tables may be perfectly plain, of

G

whitewood or pine, each on four solid legs, oiled but not varnished on top, thirty inches high (lower for schools), and eight feet long by three feet wide.[1]   It is well to have black lines ruled to mark off the territory for which each of the five students is to be held responsible. Plain chairs with rubber caps on the feet are good, though revolving chairs have advantages.   At least four feet should be left between each table to allow each student abundant room, and to permit the teacher to pass easily among them.   Shallow drawers may be made in the tables, but if many divisions of students use the same tables, these will be insufficient and may as well be omitted in favor of lockers and drawers built elsewhere in the room in sufficient number to allow one to each student.   For elementary students a drawer eighteen by twelve inches is ample, but each student should have one to himself for his tools, note-books, etc.   Stone jars under each table for waste materials are desirable.

Of other furniture, I would place next a Wardian case, or miniature greenhouse, in which plants may be kept alive while under observation or experiment.   In fact, not much physiological work is possible without something of this kind, for the dryness, gases, and other disturbances of an open schoolroom produce abnormal results, and often no results at all.   The

---

[1] Detailed descriptions and figures of various forms of laboratory tables may be found in the *Journal of Applied Microscopy* (Rochester, N.Y.), for April, 1899.

ideal place for this work is a small greenhouse opening off the laboratory; and often some angle or gable
of the building offers a place for it.   In such a house
not only could experiments and observations extending
over a considerable time be carried on, but a small
collection of typical plants to illustrate ecological principles could be kept to obvious advantage.   In place
of this, where very large, especially bow, windows are
available, a glass partition could be used to make a
small greenhouse in the laboratory; but the heating
might offer difficulty.[1]   But a simple Wardian case is
always a possibility either in laboratory or schoolroom.
Its chief qualities are abundant light, hence as much
glass and as little frame as possible, sufficient tightness
of construction to hold moisture and exclude most of
the gases and dust of the room, and some provision
for heating in case the temperature of the room falls
below about 10° C. at night, or when high temperatures
are needed for special experiments.   Such a case,[2] built
entirely of glass and metal, in use in my own laboratory, is shown in outline in Fig. 3.   The floor is a
copper box four inches deep filled with water and heated
from below by a Koch safety gas burner, whose flame
is shielded from draughts by a sheet-iron hood.   The

[1] Valuable hints upon the management of such window gardens, and
suggestions as to the best plants for them, are given by J. W. Harshberger
in *Education*, XVIII, 1898.

[2] Made for me by Williams, Brown, and Earle, of Philadelphia.

height of the flame is controlled by a Reichert thermo-
regulator inside the case, which can be set at any desired point, and which
keeps the temperature
within 3° C. of that
point, no matter how
low it falls in the room
outside.   This case is,
however, more elab-
orate than necessary,
and, after my expe-
rience with it, I believe
one would work well
if built in the follow-
ing manner.   Have
made a covered, gal-
vanized-iron box the
length of the window,
two feet wide and
three inches deep, with
a hole in one corner
for filling, and a tight
sheet - iron hood be-
neath to shield the
flame and keep gases

FIG. 3.—A successful Wardian case. Scale,
¼ inch = 1 foot.

from rising through
the joints of the case; have sashes made with as little
wood as possible to surround and cover it, as shown

in the cross-section in
Fig. 4; on one of the
long sides two doors
must be left, which
can be tightly closed;
the top may be hinged
to allow opening for
ventilation at times;
support the whole on
a firm table; shelves
of glass or wire net-
ting may be added;
use preferably a Koch
safety burner (which
shuts off the gas if the
flame goes out) and
a Reichert regula-
tor. The entire cost
should not exceed $25
to $30. It should
not be built into
the window, at all
events not without an
extra sash some in-
ches from the win-
dow sash.

Other necessary or
desirable furniture is

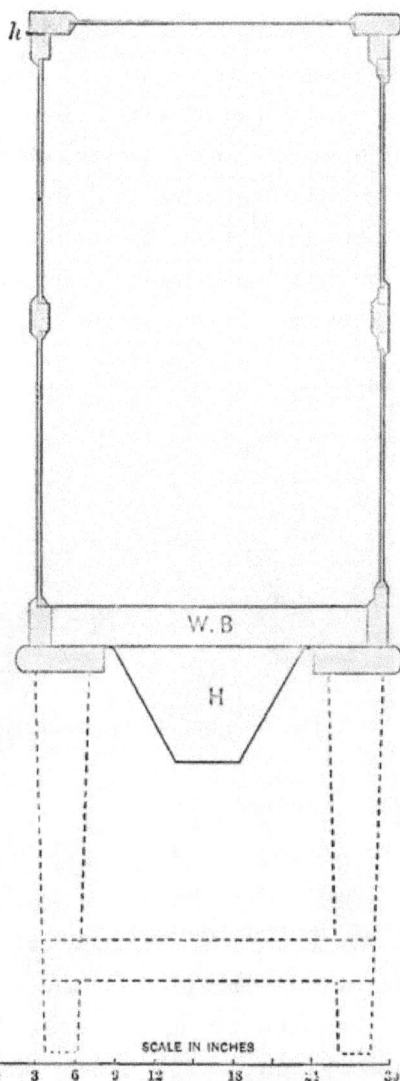

FIG. 4.—Plan for a Wardian case, in cross-section. *H*, sheet-iron hood; *h*, hinge of top; *W.B.*, water box of galvanized iron; dotted lines show legs of table.

the following. There should be one or two large tables for holding the supply of material for the class and for demonstration, etc. These may be built three feet high, with lockers for microscopes, or other storage, beneath. A teacher's platform with a blackboard is essential, and over it, as well as elsewhere in the room, should be racks for displaying diagrams. The best racks I know of are boards an inch thick, four inches wide,

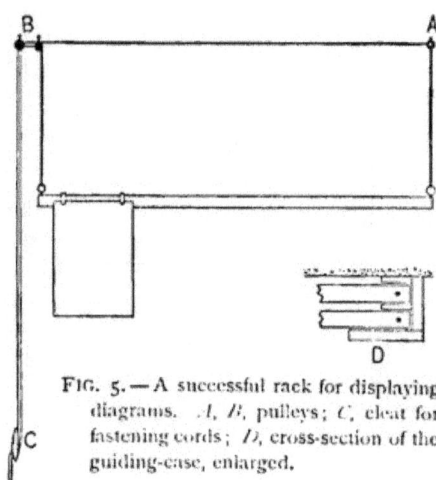

FIG. 5. — A successful rack for displaying diagrams. *A*, *B*, pulleys; *C*, cleat for fastening cords; *D*, cross-section of the guiding-case, enlarged.

and ten feet long, rounded on one edge to hold Dennison's No. 12 Card Holders (which are far the best diagram holders I have ever seen); these boards run in a light guiding frame, like a window frame, and are raised and lowered by cords attached as shown in Fig. 5, which also shows a cross-section of the guiding-case. The latter, however, is not indispensable. Two boards may be used in the same case, passing one another and giving two tiers of diagrams if the ceiling is high enough either above the blackboard, or elsewhere. A gas table for heating, glass-bending, etc., is necessary, as is a large sink (preferably porcelain-

lined) with several taps, and cases with glass fronts for
storing museum specimens, materials, etc.   Lockers
or drawer cases, when built away from the wall (as
in Fig. 1), should not be over four feet high, in order
not to obstruct a free view around the room.[1]

Of instruments the first in importance are the scalpel, two needles in handles, forceps, and hand lens,
which should be supplied as a loan to each student,
together with a box to keep them in.   These may

FIG. 6.— A successful set of dissecting instruments, with case.

be bought in various forms and qualities at prices
from 75 cents upward per set from any of the firms
mentioned later.   For use with my own classes I
have designed the set, together with their leatherette
case, figured herewith (Fig. 6), which is manufactured
for me at $1.20 each by Williams, Brown, and Earle,
of Philadelphia, and which has proved very satisfactory.
It includes all instruments essential in elementary work.

[1] There are valuable hints upon these points, and upon other matters
connected with laboratories, in a fully illustrated article on " Representative American Laboratories," in the *Journal of Applied Microscopy*
(Rochester, N.Y.), Vol. I, 1898, pp. 22–32.

The lens fits in at one end and the needles, etc., at the
other; and the case is intended either for keeping the
tools at the laboratory or for carrying them in the field.
A fair dissecting microscope may be made by placing
the lens open on the case, with the lenses, held in
position by one of the flaps, projecting over the side.
Next in importance are dissecting microscopes, which
are of the greatest value. There should be at least
one to a seat, preferably one to a student. There is
a large variety of these by different makers, and of
great range of excellence and cost. For a cheaper
kind, the Barnes Dissecting Microscope offered by
Bausch and Lomb to schools and colleges at $1.88 to
$2.82 is excellent, and, in my opinion, ample for ele-
mentary courses.

The compound microscope is the chief tool of the
biologist and indispensable to the biological laboratory.
The ideal arrangement provides one for each student;
after that, one to each seat where more than one
division uses a room; after that, one to as few students as possible. If there is one to each student, it
is easy to hold him responsible for its condition; and
its life is so much longer that it pays in the end to
provide the greater number at the start. There are
all grades of compound microscopes and all prices.
The makers best known in this country are Zeiss, of
Jena (Germany), Leitz, of Wetzlar (Germany), Reichert,
of Vienna (Austria), and Bausch and Lomb, of Roches-

ter (New York). After much experiment with many
makes, I have, for the use of my own classes, fixed
upon the instrument shown in the accompanying cut
(Fig. 7), which has been specially made by Reichert,
and is supplied, duty free, at $27, by Richards and

FIG. 7. — A successful student's microscope.

Company, of New York. Its points of excellence
are, — the very firm base, the presence of a nose-
piece (a most valuable time-saver), and a case in which
it can be kept without closing the tubes. It has
objectives 3 and 7 and oculars II and IV. It is

nearer the ideal student's microscope than any other
that I know of. If a less expensive one is necessary,
the nose-piece can be omitted, and there are other
stands and combinations by the same maker. Instruments of corresponding power are supplied by other
makers. Those of Zeiss are usually considered the
best of all, but they are also most expensive. Those
of Leitz are thought by many to offer a good resultant
between cost and quality. All foreign makes of microscopes and other instruments may be imported by colleges and schools free of duty, and to meet this the
chief American firm, Bausch and Lomb, offer special
discounts from their list prices to those institutions.
It does not pay to buy a cheap microscope, and
nothing less than a firm stand of the continental pat-
tern, with two objectives, two-thirds and one-sixth
inch focus, and two eye-pieces, should be accepted,
and a nose-piece is well worth its cost. For such
an instrument $20 or more must be paid. It is
better to have a few of this grade than more of a
poorer sort, and in buying from any other than the
firms of recognized worth, it is better to seek the
advice of some specialist. In the laboratory the microscopes should be kept in lockers, especially if there
is one to each student, or they may be kept on the
laboratory tables under glass bell-jars, or even in
their cases when there is but one to a seat. Like
other laboratory apparatus, they should be loaned for

the term to the students, who should be held fully responsible for their good condition.

In an elementary course few reagents are used, and these so rarely that it is better to place them on the tables only when needed. The best reagent bottles known to me are those in which a pipette forms the ground glass stopper, as shown in the accompanying figure (Fig. 8). These bottles and other dishes and miscellaneous glassware are happily inexpensive, and may be obtained from any of the firms dealing in chemical supplies.

Apparatus for physiological experiments must partly be made to order from directions given in books, and partly bought; and most of the needed supplies may be purchased from the firms mentioned below. Most of the articles are not used up, but once obtained are

FIG. 8. — An excellent form of reagent bottle.

valuable year after year. There is no firm known to me which makes a specialty of apparatus for plant physiology, though no doubt, in view of the rapidly increasing attention given to this subject, that lack will soon be supplied; and it will not be long before the apparatus necessary for a standard set of physiological experiments for an elementary course in Botany will be offered for sale at a fair cost, precisely as such apparatus is now offered for the Harvard Entrance Requirement in Physics.

Abundant materials in proper condition are a necessity for good study, and fortunately these are not expensive. They are partly to be bought in the markets or from greenhouses, partly collected the summer before, while, as a last resort, some of the more special materials may be bought from a botanical supply company. If the teacher has at command his own greenhouse and gardener, as many colleges have, he is fortunate. If he is near a botanic garden, he will find the director ready to aid him in anything which advances botanical knowledge. Commercial greenhouses, happily, are everywhere, and the teacher should make friends, and a bargain in advance, with the gardener for such materials as he needs, — bulbs, flowers, leaves, plants for experiment, and also for keeping certain illustrative water plants, etc. All this, together with many other incidental expenses about a laboratory, necessitates some regular income. In colleges this is generally supplied by the laboratory fee paid by the students, amounting on an average annually to about $5 for each student, which is ample. Since the public school system does not allow of such a source of revenue, its place must be taken by annual grants from the school committees, and the teacher should insist upon that sum or as near it as possible. The materials to be collected in summer (unless for the herbarium) are best preserved in glass preserve jars in water to which two per cent of formaline (also called formaldehyd and formalose) has been added; this will perfectly preserve all vegetable tissues, but since its fumes irritate the eyes and throat, the materials should be well washed in water just before they are used. It is well in the spring to go through the outlines for the next year's work and list the materials needed, as a guide for the summer collecting. Pressed flowers are sometimes recommended for study, but they are difficult for and repellent to the beginner, who should have fresh ones only.

The charts, museum specimens, and other desirable illustrative parts of a laboratory equipment are discussed in the next chapter (Chapter VI).

The only firm in the United States which professes to deal exclusively in botanical supplies, and as well to supply everything needed by botanists, is the Cambridge Botanical Supply Company, of Cambridge, Mass. A new firm, the Ithaca Botanical Supply Company, Ithaca, N.Y., has lately been organized. The Knott Scientific Apparatus Company, of Boston, Williams, Brown, and Earle, of Philadelphia, Richard Kny and Company, of New York, all deal in smaller

botanical supplies, and the names of other firms elsewhere may be found in the advertising pages of the
Botanical Gazette. For larger glassware of all sorts,
Eimer and Amend, of New York, and Richards and
Company, of New York, are the firms I know best,
but all large dealers in chemical supplies keep such
apparatus. If a large quantity is wanted, it pays
to import it duty free through some of these firms,
in which case orders must be placed two or three
months in advance, but this is not worth while for
very small orders.

VI. ON BOTANICAL COLLECTIONS AND OTHER ILLUSTRATIONS

THE only true foundation for biological knowledge is laboratory or other practical study. This method, however, has an inherent defect in that, consisting as it must in the investigation of more or less isolated topics or types, the view it gives of the plant world is discontinuous and poor in perspective. To realize the full value of the study, these types need to be correlated and located in the general system, thus contributing to the formation of one complete and correct conception. To this end, reading, lectures, and other formal instruction are of great aid, and these I have treated elsewhere in this work; but equally valuable is that comprehensive survey of a large series of forms which is made possible only by collections of living plants, of museum specimens, of photographs or charts, of models, etc. The study of these collections alone would have little meaning, but every type thoroughly studied in the laboratory becomes a centre of illumination for a zone of related topics, which have a vivid significance and interest entirely lacking without such study.

By far the most valuable of all botanical illustrations

are living plants growing untouched in their native homes. But practically the use of these' is very limited, for some of the most instructive are tropical or of other lands, and the native ones are not only often distant, especially from students in cities, but in our climate are unavailable for most of the school year. These draw-backs are partially overcome by botanic gardens, which not only bring plants together from the uttermost parts of the earth, but group them in a manner which is itself instructive. Plants in gardens, however, while valuable for investigations upon structure and classification, are nearly valueless for studies upon their natural relations to their surroundings, though they may be so grouped as to form valuable illustrations of some well-known principles of ecology. The teacher who is so fortunate as to be within reach of a botanic garden should make the acquaintance of the director and obtain permission for himself and his students to use it freely, which will usually be readily granted. Botanic gardens are very numerous in Europe, but rarer in this country; the principal ones of North America are the following, arranged in order from east to west : —

The Arnold Arboretum (a department of Harvard University), at Jamaica Plain, Mass.

The Botanic Garden of Harvard University, at Cambridge, Mass.

The Botanic Garden of Smith College, at Northampton, Mass.

The Botanic Garden of McGill University, at Montreal, Canada.

The New York Botanical Garden, at New York City.

The Botanic Garden of the University of Pennsylvania, at Philadelphia, Pa.

The Botanic Gardens of the United States Department of Agriculture, at Washington, D.C.

The Buffalo Botanical Garden, at Buffalo, N.Y.

The Botanic Garden of the Michigan Agricultural College, near Ann Arbor, Mich.

The Missouri Botanical Garden, at St. Louis, Mo.

The Botanic Garden of the University of California, Berkeley, Cal.

There are others also, but of less complete organization, in connection with some other colleges, especially some of the State and agricultural colleges. By far the most important of the above list are the Arnold Arboretum, the Missouri and the New York gardens (the latter now forming), next to which comes that of Harvard University. The Smith College Garden was especially planned from the start as a teaching garden, and as such is fairly complete.[1] School gardens have scarcely at all received attention in this country, but a noteworthy article by

[1] A full account of it, with a plan, is contained in *Garden and Forest*, Vol. X, p. 512, 1897. The New York Garden is described in the three Bulletins of the garden by Dr. N. L. Britton, an important address by whom, on "Botanic Gardens," is in *Garden and Forest*, Vol. IX, p. 352. The Michigan Agricultural College Garden is described by W. J. Beal in *Garden and Forest*, Vol. VIII, pp. 303, 322.

H

H. L. Clapp, on "School Gardens" in the *Popular Science
Monthly* for February, 1898, gives a description of a
remarkably successful garden on the grounds of a
Boston school, and shows how much may be done with
limited space and means.   Such gardens must repay
many fold their cost, not only in botanical instruction,
but in moral influence, and their formation cannot be too
highly commended.   There are very practical directions
upon this subject in L. H. Bailey's "Lessons with
Plants," and especially in his "Garden Making," showing how much can be done at little or no expense, and
a recent book in German, illustrated with plans,[1] is
devoted entirely to this subject.

An essential feature, indeed the most essential feature, of all botanic gardens are their ranges of greenhouses, for thus are the living plants made independent
of climate and country.   Such collections illustrate
extremely well most structural features, and fairly well
many ecological principles, especially where the natural
conditions are carefully imitated, as is to some extent
possible with water plants, desert plants, epiphytes, etc.
If the teacher has not the use of such a collection, and
has no school greenhouse, he can perhaps make the
acquaintance of some owner of a private or even a
commercial greenhouse and persuade the owner to
accumulate some of the more important forms.   What

[1] Cronberger, B. "Der Schulgarten des In- und Auslandes." Frank-
furt a. M. 1898. 2.80 marks.

these forms are, he will know from his own earlier studies.

Next in value to living plants come dead ones preserved to look as much like life as possible. The collection and arrangement of such specimens is the function of museums. Unhappily, there is no known method of preserving plants in their natural forms and colors as is possible with so many animals; though on the other hand it is possible to preserve plants, when dried, with cheapness, compactness, and accessibility far exceeding what is possible with animals. Hence it comes about that there are many great herbaria and but few great botanical museums. Even in Europe botanical museums are very scarce and of minor interest, and in America there is as yet but a single botanical museum of any account, that of Harvard University, and this one owes its interest chiefly to the success with which the living plants, including flowers, have been imitated by glass models of the most natural form, size, and color.[1] In time the New York Botanical Garden will undoubtedly possess a museum of the greatest comprehensiveness and value. Most colleges with depart-

---

[1] This collection of models is being made by Leopold and Rudolph Blaschka, of Dresden, Germany. It has attracted wide attention for its great accuracy of execution. A full account of it is given by Walter Deane, in the *Botanical Gazette*, XIX, p. 144. One of the curators of the British Museum has said of it, "No other museum possesses anything half so beautiful." It is unique, and by contract with the makers no part is to be duplicated.

ments of Botany possess small teaching collections, and these are of such value that every teacher of an elementary course should aim to gather at least a small museum.

Many parts of plants, such as hard fruits, woody stems, etc., may best be preserved dry, as indeed may the entire plants themselves, in herbaria, of which I shall speak presently. But the softer parts can be kept only in some preservative liquid, though none is known which will keep color well. A solution of three per cent formaline in water will preserve color as well as any, but it keeps some colors much better than others; and in it the important green tissues become of a translucent unnatural shade, which is hardly worth the having.[1] In my own collections I use a mixture of two per cent formaline in thirty per cent alcohol, which preserves the softest tissues perfectly in every respect except color. Formaline is used in such small quantities that it is really very cheap; and colleges and schools are entitled by law to purchase, though with rather complicated legal formalities, alcohol free of internal revenue tax, which makes it cost only about 40 cents per gallon in quantity. Bottles for specimens may be any one of the many forms of preserve jars; but after considerable experience with their effect

[1] If one wishes to try to preserve the green color by other methods, he may consult to advantage an article by A. F. Woods, in the *Botanical Gazette*, XXIV, p. 206.

upon classes, I am convinced that it is true economy to
buy only the best white flint glass bottles with ground
glass stoppers, not only for specimens in liquids, but
also for dry objects, such as seeds, which need some
kind of a vessel. Thus not only are all specimens safe
from evaporation and dust, but the respect of the student is far greater for a compact, artistically presented
specimen than for one in a green leaky jar or a dusty
box, and hence its value to him is greater. The
teacher, too, is more likely to accumulate only things
of value if the receptacles must be economized. For
a collection of my own I prefer a dozen such specimens
to thrice that number indifferently prepared. I have
experimented with several forms of bottles, and finally
have fixed upon Whitall and Tatum's (Boston and New
York) No. 2605 specimen jars, which may be had in
all sizes, and for which their published prices are subject to large discounts. I prefer the appearance of
these to that of the kinds without a neck. But of
course if one cannot afford such bottles, some of the
many forms of preserve jars will do very well, and are
far better than nothing at all. In whatever manner
prepared, however, every specimen should be in condition to be handled and passed about. Tight, upright,
glass-fronted wall cases should be provided for them,
and it is well to have them very fully labelled and carefully arranged upon a definite plan in order that they
may be as instructive as possible when not actually in

class use. In any museum collection whatever, the great guiding principle should be selection, not accumulation; and in plan and labelling the famous dictum of Goode[1] should be remembered, that the modern museum is a collection of labels illustrated by specimens. The teaching collection need have no formal beginning; but as specimens from one source and another are obtained, they should be properly prepared and added. There are as yet no firms offering for sale considerable numbers of museum specimens of plants such as are offered of animals.

It is of the greatest importance, however, that the collection should grow upon some definite plan, as otherwise half of its value is lost. One may, according to his tastes or facilities, take as the leading idea the illustration of the principles, either of morphology, ecology, or the natural groups. An ideally complete collection would include all three. Following is a suggestion for a plan based upon that which I have worked out for the collections under my charge :—

[1] There are valuable papers on Museum-making, by G. Brown Goode in *Science*, New Series, Vol. II, p. 197, and Vol. III, p. 154. Particularly apposite and most valuable, though I cannot agree with all of its recommendations, is J. M. Macfarlane's "The Organization of Botanical Museums for Schools, Colleges, and Universities," in *Woods Holl Biological Lectures for 1894*. Of much suggestiveness is Boyd Dawkins' address on the Place of Museums, in *Nature*, July, 1892, p. 280. See also *Nature*, 1895, p. 107.

Division I.

Morphology. *A.* Phylogeny of the Plant Kingdom ; progress from thallus to shoot and root ; and from sporangia to ovules and anthers.

B. Particular anatomy and morphology of Thallophytes, Bryophytes, and Pteridophytes.

C. Particular anatomy and morphology of Spermatophytes.

    1. The root, typical form and plasticity.

    2. The shoot.

        a. Stem, typical form and plasticity.

        b. Leaf, typical form and plasticity.

        c. Flower, typical form and plasticity.

        d. Fruit, typical form and plasticity.

Division II.

Ecology. Adaptations connected with particular or typical modes of, —

A. Nutrition.

    a. Absorption.

    b. Transfer, including transpiration.

    c. Metabolism, including photosynthesis.

    d. Storage.

    c. Secretion and excretion.

B. Growth.

C. Reproduction.

D. Irritability, *i.c.* individual response to external stimuli.

   E. Locomotion.
     a. Of pollen.
     b. Of seeds and some vegetative parts.
   F. Protection.
     a. Against weather conditions.
     b. Against living enemies.
Division III.
  The Natural Groups of Plants. (Natural History of Plants.)
    A. The Algæ.
    B. The Fungi.
    C. The Lichens.
    D. The Bryophytes.
    E. The Pteridophytes.
    F. The Spermatophytes.
      I. Floristic Divisions.
      II. Ecological Divisions.
        Mesophytes.
        Hydrophytes.
        Xerophytes.
        Halophytes.
        Climbers.
        Epiphytes.
        Parasites.
        Insectivora.
        Myrmecophila.
        Etc.

Of course this plan is too comprehensive to be carried out in its entirety in a teaching collection, and it is offered but as a suggestion. In a large public museum, other sections, to illustrate palæontology and economics, would be added, together with the fullest representation of all phases of the subject by models, paintings, photographs, apparatus used in investigation, etc. Even in the smallest collection there should be the fullest labelling, which should give the exact place of the specimen in the plan. As an example, I give here a typical label as adopted for my own collection (Fig. 9). These labels need not be permanently attached to the bottles, but are to be placed with them when not in use by the class,

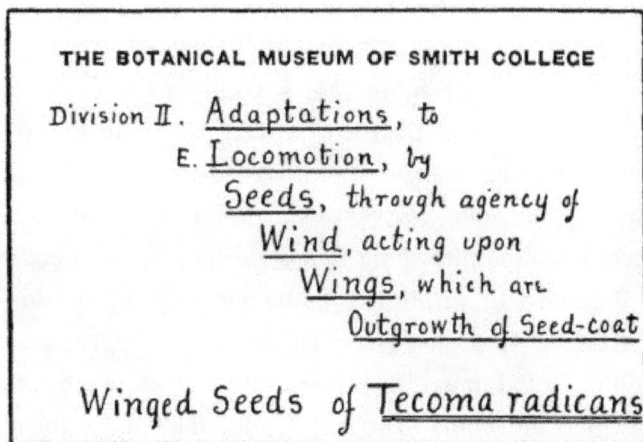

THE BOTANICAL MUSEUM OF SMITH COLLEGE

Division II. Adaptations, to
  E. Locomotion, by
      Seeds, through agency of
        Wind, acting upon
          Wings, which are
            Outgrowth of Seed-coat

Winged Seeds of Tecoma radicans

FIG. 9. — Sample museum label.

though a briefer label should also be kept in each bottle. It is well to have the museum specimens always visible and accessible.

Theoretically, an herbarium is a part of a botanical museum, but on account of its special nature and use it is kept stored by itself and not on exhibition, though sometimes a few pressed plants are exhibited behind

glass, like pictures.   For investigation into and illustration of systematic Botany, an herbarium is absolutely
indispensable; but in a teaching collection, for use where
the work is not primarily systematic, the plan of the
collection should accord with the plan of the teaching.
The question now arises whether it is not possible to
utilize the great ease, cheapness, and compactness of the
herbarium method of preservation of plants in the
formation of a collection to illustrate the principles of
morphology, ecology, and natural history.   I have experimented not a little upon this subject to the conclusion
that an herbarium of the greatest usefulness can be
made upon the same plan as is outlined above for a
botanical museum.   Every specimen in it would be
selected to illustrate some fact or principle, and need not
at all consist of an entire plant, but only the portion of
it useful for this purpose.   Drawings, photographs, full
labelling, etc., may be incorporated in it much easier
than in a museum.   Indeed, I am inclined to question
whether, in cases where means and room are very limited, the herbarium on this plan may not be superior to
the museum.   It must, however, always suffer the drawback of not being constantly visible to all, though even
this might be overcome by keeping the sheets mounted
in glass-fronted frames, like pictures.   For use with a
class the sheets would temporarily be placed in glass-
fronted frames with removable backs.   Of course the
ordinary herbarium methods would apply to the prepa-

ration of specimens for such a collection.[1]   It may be
that such an herbarium would find its highest usefulness
as a private collection, built up to illustrate his own
studies by the teacher, or by his best students.   To illustrate the possibilities of the plan, there are here added
photographs of two sheets from my own collection, one
ecological and one morphological (Fig. 10).[2]

There is yet another important phase of herbarium-
making in elementary teaching.   Many teachers are
accustomed to require from their students the making of
one of a definite size as an integral and important part
of their courses.   There are many conditions under
which this plan seems to me of value, as when facilities
for any other actual work with plants are entirely wanting, or when overworked or undertrained teachers can
give the science but scanty attention; and certainly it
gives opportunity for careful manual work which always
has moral value.   But, viewed from the broader educational standpoint, the requirement of an herbarium from
elementary students seems to me quite uneconomical, in

[1] The fullest account of herbarium methods is contained in W. W.
Bailey's " Botanical Collectors' Handbook," and in Chapter X, Section IV,
of Gray's " Structural Botany"; there is valuable matter also in the Herbarium Number of the *Botanical Gazette* (June, 1886); in Mr. Walter
Deane's series of five articles " Notes from my Herbarium " in the *Botanical
Gazette,* Vols. XX and XXI; and in L. H. Bailey's " Lessons with Plants,"
p. 437.   On preserving colors in dried flowers, there is a valuable note
in " Annals of Botany," Vol. I, p. 178.

[2] These sheets but partly illustrate the value of the plan, as they were
prepared as an experiment before it was fully worked out.

FIG. 10. — Photographs of two sheets from a small morphological and ecological herbarium; × about ⅓.   In the herbarium, the uses or the nature of the parts, with the names of the plants, are written where the numbers (rendered necessary by the method of engraving) stand in the cut, and are as follows :   The left-hand sheet is devoted to *Morphology, Stipules ;*   1, stipules as part of foliage (*Geum* sp., probably);   2, bud-coverings (*Humulus lupulus*) ;   3, spines (*Euphorbia splendens*) ;   4, part of foliage (*Galium*) ;   5, all of foliage (*Lathyrus Aphaca*);   6, spines, also dwellings of protecting ants (*Acacia sphærocephala*) ;   7, bud-coverings (*Passiflora*);   8, tendrils (*Smilax*) ;   9, bud-coverings (*Liriodendron Tulipifera*).   The right-hand sheet is devoted to *Ecology, Climbing Organs ;*   10, axis of compound leaf (*Bignonia*);   11, hooked epidermal spines (*Rosa sinica*);   12, aerial roots (*Ficus repens*);   13, axis of leaf (*Lathyrus Aphaca*);   14, axillary branch (*Passiflora*);   15, petioles (*Clematis Virginiana*);   16, main stem, twiner (*Aristolochia*);   17, extra-axillary branch (*Ampelopsis Veitchii*);   18, stipules (?) (*Smilax*).

that the labor necessary for collecting, drying, and mounting the specimens is largely not botanical, and is excessive in proportion to the amount learned through it about plants. The question before us in such cases is not whether a thing is valuable or not, but rather, what will yield the largest returns for the time and energy expended. Moreover, the great majority of people have no taste for collecting, and extremely few ever keep it up; so their school labors in this direction result in a bulky pile difficult to store and unattractive to preserve. It is not, I would repeat, that such herbarium-making has in it no profit; but simply that it is not as a whole profitable. On the other hand, under some circumstances, it may be very valuable, as when it is used to cultivate in those with a talent for natural history the collecting instinct, that first and plainest mark of the naturalist. The best plan, then, would seem to be to make the collecting voluntary, to be taken up by those whom it interests.

As to the plan of such an herbarium, what has already been said about the museum herbarium applies here with equal force, and I believe it may best represent, not the flora of a region, but principles of morphology and adaptation. I have found in my own experience that this plan interests many who care not at all for a floristic collection. The search in the native flora for examples of the different forms of stipules, for kinds of protective structures, illustrations of marked adaptations to special habits, etc., must surely have a zest not inferior to the gathering of all the species of a given area, though this also is not to be disparaged. Moreover, a collection of this kind has a practicable limit of completeness, which a floristic one hardly has, and a small one expresses more than a floristic one of the same size. There are other plans on which herbaria may profitably be made. Professor L. H. Bailey, in his " Lessons with Plants " (pp. 443–444), recommends special collections and suggests various sorts.

In making these students' herbaria, most teachers require the standard size of mounting paper, the regular genus covers, etc. But while this size ($16\frac{1}{2} \times 11\frac{3}{4}$ inches) is very convenient in large herbaria with proper cases, one or two hundred of such sheets make a package very awkward to store amongst a student's other effects, and not easy to consult on small crowded tables. This objection can be overcome if the sheets can be reduced to the size of a large book and kept stored among books. This I have found to be entirely practicable, and so advantageous that I have adopted it for a small private collection of my own, even in the presence of the best facilities for storing the larger size. Sheets one-half the usual size will hold most specimens (see Fig. 10), and those too large can be treated precisely as are those too large for the ordinary size of sheets. The speci-

mens are firmly glued to the half or somewhat smaller
sheets, which are then placed between those covers
used in colleges by students for holding any number of sheets of paper, and held by paper fasteners.
The thickness of the specimens is compensated by
extra strips or stubs, and additions and rearrange-
ments may be made with great ease. The collection is then practically a book, and may be kept
among books. The specimens are amply large for
amateur's use. If it be thought that specimens so
kept are particularly liable to dust and insect-rav-
ages, it must be remembered that they are no more
so than they are in the usual condition in which beginners keep them, and that if one cares, he may
keep these books also in tight tin cases. It is sometimes said in favor of the standard size that if a student continues his studies, his collection will form a
nucleus for his larger herbarium; but it is not fair
to put the dozens who go no further to much incon-
venience for the sake of the rare one who does.

Specimens of the plants themselves include also the
various anatomical preparations, skeletons to show the
fibro-vascular system, wood-sections, etc., and particularly microscopical preparations. It is well to have a
wide range of the latter for demonstration and for
voluntary study by those whose tastes incline them to
it; and it is also profitable to have some sets for use
in the regular class work, as recommended in certain

places in the outlines in this book. These sets may be bought from various dealers in microscopical supplies, but are better made from time to time by the teacher himself, or by the specialists among his pupils. By the addition of a few each year a valuable collection will soon result. Proper cases, of many forms and prices, for storing them are supplied by dealers.

Next in illustrative value after preparations of the plants themselves would come, theoretically, good models of them; but practically I find good pictures or diagrams better, and shall treat these first. Of pictures, as a rule, photographs are best; and where comprehensive or complicated things are to be shown, or where the actual living form and surroundings are important, they are indispensable. Their chief drawback is that when large enough to be shown to a class they are very expensive. This can be overcome by photographing them on glass and projecting them to any desired size on a screen by the well-known stereopticon method. This, however, is of little use in connection with laboratory work and is most useful where lectures are a part of the mode of instruction. Good photographs thus appealing vividly to the mind through the eye seem to me of the greatest value, especially for ecological studies (where, indeed, they are invaluable), and for representing the natural appearance of many important plants from foreign parts which grow badly or not at all in greenhouses, or for

showing the general topography of masses of vegetation. A standard collection, selected by a specialist,
of lantern slides of this character which could be
purchased in one set, would be of great value and
doubtless will soon be offered by some of the dealers,
who already offer heterogeneous lots.[1] The best
stereopticon is one using the arc electric light; it is
handier, cheaper, and better than the calcium and
other forms, and is so powerful as to need no elaborate
system of dark shades to the room, but may be used in
almost full daylight. Many forms of such lanterns are
offered; I use to my satisfaction one made by J. C.
Colt, of New York. The best screen is a smooth white
wall. With such a lantern, and an ordinary microscope,
one may also project microscopic objects upon a
small screen, and thus show tissues, circulating protoplasm in Nitella, etc.; but in general the manipulation
is so time-consuming and difficult that it is hardly
worth while unless one has a liking for that very
kind of thing. Where a lantern is impracticable, it
is still decidedly worth while to collect photographs, of
which many of value are now obtainable from return-
ing tourists and other sources. A superb collection
has recently been made available in Schimper's new

[1] A great number indeed, selected by Koch, is published by Krüss in
Hamburg; a catalogue may be obtained from any dealer in botanical
supplies. Unfortunately, these are mostly but woodcuts from books, not
photographs from nature. Most American dealers in stereopticons also
offer botanical slides.

I

work "Pflanzengeographie," on which further information is given in the next chapter. Large photographs of microscopic sections have some value, and an unusually fine series, by Tower, is sold by Ginn and Company, Boston.

After photographs, and for some purposes before them, come drawings or diagrams, which are of two general sorts, — those intended to show the very living appearance of plants or their parts, including enlarged views of small organisms, and those intended merely to help to a vivid vizualization of their structure. The former are not of much value unless very well done, true in perspective, and correct in coloring; indeed, it may be said their value is in direct proportion to their artistic excellence. Good examples occur among the diagrams of Kny, Dodel, and Peter, to be mentioned below. Not much can be done toward making home-made diagrams of this kind unless an artist is available. In the second kind, however, those to simply help the mind to form a three-dimensioned conception of some complicated structure, the element of objective correctness is not so important; and of such diagrams I think the very best is that which grows before the student's eyes on a blackboard under the hands of a teacher, with copious explanation and the aid of colored crayons, etc. Some skill in blackboard drawing is very desirable in the teacher, and I have no doubt that in time the increasing care devoted to the education of teachers of

Botany will lead to their instruction in this useful art. Next to such diagrams, and for some purposes superior to them, are the excellent published diagrams of anatomy in the Kny and in the Dodel series. The best of these is that of Kny, with full explanations in German, of which one hundred have appeared, each 84 × 68 cm., costing about $85 for the set, and the teacher should make every effort to obtain this series. They should be mounted on cloth for greater resistance to wear. Another valuable series is that of Frank and Tschirch, devoted to the physiological aspects of structure, sixty in number, of the same size as Kny's and also with explanations in German. The series of Laurent and Errera, fifteen in number, slightly larger than those of Kny, with explanations in German, French, and English, is also good. If one cannot afford the Frank series, the latter is a fair substitute. The Dodel-Port series, slightly larger than the Kny series, is also valuable, as is the Peter series, published by Fischer in Berlin. Full particulars of these may be obtained through any of the dealers in botanical supplies or in foreign books.

Many teachers value diagrams made by themselves or students above these printed kinds, holding that they have much more meaning, and hence value, and are also much cheaper. There are several methods of making them. One of the simplest and best is the use of strong, light-brown manila paper for the back-

ground, and India ink, applied with a brush, for the
lines. This method is inexpensive, easy, and gives a
pleasing combination. If one wishes to use colors,
water-colors are best; but a fair substitute may be
found in colored crayons, which may be prevented
from rubbing by a previous immersion in melted soft
paraffin until the bubbles cease to come off, or by
spraying the drawing through an atomizer with a weak
solution of gum arabic.[1]

The best method known to me of hanging diagrams
when in use has already been described (see Fig. 5).
For storing they should be as nearly as possible of one
size, of which the Kny series is a standard (84 × 68 cm.).
They may then be very conveniently kept in shallow,
upright cases built against the wall, a foot above the
floor, with the front hinged on the bottom so as to drop
forward a few inches at the top, as shown in the accompanying diagrammatic cross-section; a chain keeps the
front from falling too far (Fig. 11).

Another very valuable class of illustrations includes
those in special monographs or technical papers; and
where a good library is available, free use should be
made of these original sources of information.

Many teachers would probably place models before
diagrams in illustrative value, and chiefly because these,

---

[1] Another method is fully described, and there are other valuable
hints upon this subject, in "Natural History Charts and Illustrations,"
by J. W. Harshberger, in *Education*, April, 1897.

like the objects they represent, are of three, not two, dimensions. The latter point is, I think, of more theoretical than practical importance; and proper perspective in drawings, and especially in photographs, gives the same result. Moreover, models are far more difficult and expensive to construct with accuracy and truth to nature than are drawings, and this applies particularly to minutiæ of structure. Botanical models are generally made of papier-mâché or gelatine, sometimes of wax or glass. Enlarged models of flowers or other parts, which are familiar to everybody in a living condition, seem such a grotesque parody of nature that they inspire more amusement than respect in the student, especially in those kinds made to come apart to show what is concealed within. To use such elaborate methods to illustrate

FIG. 11. — A successful box for storage of diagrams, in cross-section. The dotted lines show it open. Scale, about 1 inch = 1 foot.

facts which any one, with aid of a knife, can see in a minute with his own eyes seems to be carrying the good principle of clear illustration over the bounds of the useful into the ridiculous. It certainly is pos-

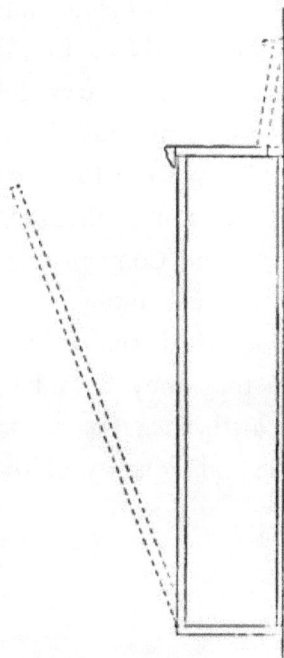

sible to refine illustration to a needless and enervating extent. These objections, however, do not apply to enlarged models of minute and difficult subjects, such as embryological development, nor to models of entirely unfamiliar objects; and especially it does not apply to such models as the Blaschka series in the Botanical Museum of Harvard University, which inspire in the beholder no sensation except wonder and admiration.

The principal makers of botanical models are Auzoux, of Paris, and Brendel, of Berlin, and in this country Kny and Company, of New York, make a specialty of their importation. There are purchasable, also, useful models of spiral vessels, of stomata made of rubber so they may be inflated, of fibro-vascular bundles in growth, etc., the usefulness of all of which varies with the individuality of the teacher.

VII. ON BOTANICAL BOOKS AND THEIR USE

Books are the storehouses of knowledge, but in order to make full use of their advantages, one must learn where and how to seek in them that which he needs. How to use books profitably is therefore an important phase of the education of both teacher and student. For the teaching botanist, books fall into three classes: first, those to be read for self-improvement; second, books of reference; third, text-books for class use.

In the preceding chapters I have tried to emphasize the real aim of scientific teaching, which is the cultivation of the scientific habit of mind to the end that a scientific instinct may become a part of the student's mentality. No teacher who lacks this scientific habit of thought, or who has it but indifferently developed, can lead others into it, and he is likely to be the most successful teacher who has it the best developed. Self-improvement in this respect is therefore a first duty of every teacher, and while the best of all ways lies through original investigation, something can be accomplished by the reading of good books, especially such as are recognized as models of scientific exposi-

tion. In reading such books, however, it will be of
little use to skim them for their facts or their rhetoric;
but the reader must minutely enter into the spirit of
the work, try to put himself into the very mental
attitude of the writer, with him view the original data,
follow him as he marshals these into their proper
relative positions, and try even to anticipate him in the
deduction of his general principles. Happily there
are many good books which will fully repay such
reading.

Upon the general subject of scientific education,
and the true place of science in education, there are
first of all the various addresses of Huxley, contained
in his Collected Essays, particularly in the volume
entitled "Science and Education." Of the greatest
importance are also the addresses of President Eliot,
now accessible in his "Educational Reform." Among
books which are models of scientific argument, I think
the first place should be given to Darwin's "Origin
of Species"; and if the teacher can thoroughly study
but one book, it should be this. Its matter has some
of it been superseded, but its spirit has not. Suggested naturally by this work are others of Darwin's,
of which, perhaps, the "Power of Movement in
Plants" would most interest the botanist. Some of
Huxley's biological essays are also not inferior to
Darwin's in scientific exposition, and are much superior in literary form, but their subjects are of less

botanical importance. There is, however, a series of botanical essays which are among the best of models, those of Dr. Asa Gray, contained in his "Scientific Writings," particularly those that relate to geographical distribution, though nearly all in the two volumes will attract and instruct the American botanist. Another model of scientific writing, a work of charming style and great force, and one that it will pay every teacher to read from cover to cover, is Sachs's "Lectures on the Physiology of Plants." Sachs's "History of Botany" is also a classic, most readable and suggestive. There is one disadvantage common to all of these works, but one unavoidable in all scientific books while the science is advancing as rapidly as at present; namely, much of their matter has been superseded by later researches. This drawback the teacher can in part compensate by reading good new works as they appear, whose standing can be judged by the reviews of them in the botanical journals.

It is, of course, of the greatest profit to the teacher to keep in touch with botanical progress through the botanical journals. The leading journal of this country is the *Botanical Gazette*, which, in addition to technical articles, gives summaries of new discoveries, reviews of new books, and many notes of general interest, though naturally most of the matter is not utilizable except by those who have had thorough college courses in botany. The *Bulletin of the Torrey Botanical Club*, another leading journal, is very special in character.   There is also much of botanical interest in *Science*, the leading American scientific journal, which every teacher of scientific subjects should certainly read regularly.   Of a much more popular character is the *Plant World;* and the *Asa Gray Bulletin* makes a special effort to provide material of value to school-teachers of elementary classes in Botany.   To teachers in New England who are studying the New England flora, *Rhodora* will be indispensable.   Particulars as to price, etc., of these journals will be found in the Bibliography at the end of this chapter.   Of course there are very numerous special botanical journals, but these mentioned are likely to be of most importance to the American teacher of elementary courses. Sample copies may be obtained of any of them from the publishers.   Where the teacher lives near a public library, the authorities can no doubt be induced to add some of these to their reading rooms, and indeed the rarer or more expensive botanical books are often obtainable in this way.

There will be a place in the teacher's reading, and in that of his students also, for books of a less special character which may be read for both instruction and entertainment.   Among such works, one of the very best is Wallace's "Malay Archipelago."   Another classic

work is Bates's "Naturalist on the Amazons," and
another is Belt's "Naturalist in Nicaragua," while for
great interest as a book of travel combined with philo-
sophical observations upon natural history, Forbes's
"Wanderings of a Naturalist in the Eastern Archipelago" must rank very high. To these I would add
two books by Hudson, "The Naturalist in La Plata"
and "Idle Days in Patagonia," which works in my
opinion are quite unmatched for their combination of
intense interest, clear scientific description, and fine
literary form. Most of these books deal, it is true,
much more with animals than with plants ; but they will
not lack interest for the botanist on that account. If
one would read a very entertaining and instructive book
in German, he should take Haberlandt's "Eine botanische Tropenreise," a book of travels in the tropics by
a botanist, a work commended to young botanists
studying German. And all young botanists should
study German, for they cannot go far in scientific study
without a knowledge of it.

A class of books very influential for good, and as yet
far too few in number, are collections of essays upon
important botanical topics, written authoritatively and
attractively. Such works are useful both to teacher and
students, and to general non-scientific readers as well,
and they may attract to the science many who would
not think of approaching it through its more scientific
phases. Such a book, warmly to be commended for its

combination of scientific spirit with attractiveness of
style, is Geddes's "Chapters in Modern Botany." Another is Sir John Lubbock's "Flowers, Fruits, and
Leaves," devoted to some of the most attractive of
ecological problems.   Another is Sargent's recent
"Corn Plants." Another, concerned mainly with physiological topics, is Arthur and MacDougal's "Properties
of Living Plants." A very modest and little-known
book of ecology is Dr. Gray's "How Plants Behave."
Of this character, too, are the chapters in Kerner and
Oliver's "Natural History of Plants," a superbly illustrated four-volume work, which is a perfect treasury of
ecological information and suggestion.   It must be used
with some caution, however, since its author is over-
sanguine at times in his discovery of adaptations where
others have not been able to see them.   But caution is
necessary in reading all books, and it is needful ever to
remember that a thing is not necessarily true because
even the best book says it is.   Another volume of
botanical essays full of interest and suggestiveness,
dealing with evolutionary topics, is Bailey's "Survival
of the Unlike," and indeed one may well bring the same
author's "Lessons with Plants" into books of this class,
especially for young people.   Such books should be in
every school library, and students allowed the freest
access to them.   The above list by no means includes
all good books of this sort, but only some of the best
of them, and all grades exist from these down to good

popular works, and through these to many that are of little or no value. I have not myself given attention to the popular books, but there are references to a few of the best in L. H. Bailey's "Lessons with Plants," p. 443.

Passing next to the important subject of reference books, I shall enumerate the best in each department of work likely to be taken up in an elementary course. Reference books have several purposes for the teaching botanist: they are sources of information when new points come up on which information is needed; they supply new methods in manipulation when new subjects are taken up; they are full of suggestions to the brighter students who take pleasure in looking through them; and they supply illustrations and additional subject-matter for fuller treatment of particular topics. They should always be accessible in the laboratory, and students should be encouraged to use them constantly, following up through the indexes the topics in which they may be interested. This habit of constantly consulting the literature is a most important one to cultivate in students. Where a school library cannot afford all of these books, but can buy some, the first mentioned under each of the classes here described should be selected. Text-books will be considered by themselves later. The following list is, of course, not intended to be exhaustive, but simply to include the most recent and authoritative works. The science is advancing so rapidly that even the best of books are soon superseded unless kept up with advances by new editions.

Upon structural botany (*i.e.* external anatomy) the work of undisputed preëminence is Gray's "Structural Botany," a very clearly written and well-illustrated work; and in condensed form the same merits prevail in his "Elements of Botany." The morphology of these works, however, is not modern in spirit, but of a formal sort, largely laid aside by modern investigation. On morphology the most authoritative work is Goebel's "Organographie der Pflanzen," now appearing in parts, a work for which there is no equivalent in English, and which, it is to be hoped, will soon be translated. There is really at present no work in English giving the results of modern studies on the morphology of the higher plants, though several of the text-books mentioned later contain it in part.

In physiology we may distinguish two classes of works, those giving practical directions for experimentation and the general hand-books or text-books. Of the former, MacDougal's "Experimental Plant Physiology" is one of the simplest and most practical works. Darwin and Acton's "Practical Physiology of Plants" (second edition) is most excellent and suggestive, while Detmer's widely used work, translated into English by Moor as "Practical Plant Physiology," combines a practical laboratory guide with a good text-book of the subject. Of general works on plant physiology, the most readable and illuminating, though one now much behind the present state of knowledge, is Sachs's " Lectures on the Physiology of Plants," a work that should be much used for the value of its point of view. The greatest work on physiology is Pfeffer's "Pflanzenphysiologie," of which Volume I has appeared, with Volume II to follow soon; the work is being translated into English, and will be indispensable to every botanical library. Another most excellent book, very direct and suggestive, is Sorauer's work, translated by Weiss as " Popular Treatise on the Physiology of Plants." Vines's "Lectures on the Physiology of Plants" is also very excellent, though now needing revision. A book much used in this country is Goodale's "Vegetable Physiology," a clear synopsis of its subject, but now also much in need of revision. Of course the various modern text-books to be mentioned later contain physiological sections.

The very important and rapidly developing subject of ecology has not yet many good works in English. First among them are several of Darwin's books, which are foundation works in some phases of ecology. Highest among general works would stand Kerner and Oliver's "Natural History of Plants," which is to be consulted with the caution already referred to. In German there are important works by Ludwig and by Wiesner. In one of its most practical and interesting aspects, namely, in the explanation of the causes of

the topography or physiognomy of vegetation, and the
distribution of the different forms, there is a most
admirable work by Warming, written in Danish, and
translated into German under the name " Oekologische
Pflanzengeographie," and now being translated into
English. More recent, and noteworthy for its great
authority, its remarkably full and clear treatment of
its subject, and its superb illustrations, is Schimper's
" Pflanzengeographie auf physiologischer Grundlage."
This work should be in every botanical library for its
illustrations alone, even if one cannot read German. It
supplies by far the best collection of botanical, especially ecological, photographs ever published. One of
the most important phases of ecology is that of the
locomotion of pollen (often wrongly called cross-fertilization); and on this there is a most valuable work by
Müller, translated into English by Thompson, under the
title " Fertilisation of Flowers." This is in great part
arranged on the dictionary principle, so that it is easy
to find out what is known of the pollination of any
particular flower. Much has been discovered since its
publication ; and a recent work by Knuth, "Handbuch
der Blütenbiologie," in two volumes with more to come,
brings the subject down to this date, at least for European plants. On the attractive subject of seed-loco-
motion there is no single large work in English, but
an excellent little book is Beal's " Seed Dispersal."
Darwin's works on fertilization of flowers, and his

"Climbing Plants," contain much ecology, and two of the most recent text-books, those by Barnes and by Atkinson, contain sections specially devoted to it.

For microscopic anatomy there is one very excellent work, full of the most practical and the most scientific information, Strasburger's "Das botanische Practicum," of which there is a condensation called "Das kleine botanische Practicum," translated by Hillhouse, under the title "Practical Botany." By aid of this book one could, without a teacher, work through a valuable course in plant anatomy; and it would be a great advantage if there were similar works for other phases of the science. On the ecological phases of the minute anatomy of plants the great work is Haberlandt's "Physiologische Pflanzenanatomie," a work that is unfortunately not translated. De Bary's work, translated by Bower and Scott as "Comparative Anatomy of the Vegetative Organs of Phanerogams and Ferns," is a standard, of great fulness and authority.

On the natural history of the groups of plants are several excellent works, of which the greatest is Engler and Prantl's "Die natürlichen Pflanzenfamilien," in German, of which twelve volumes have appeared with three or four to follow, profusely illustrated and most authoritative. Goebel's work, translated as "Outlines of Special Morphology and Classification," is very valuable, but needs revision. More recent is Warming's excellent work, translated by Potter, under the title "Systematic

K

Botany." Some of the text-books to be mentioned
below, notably Strasburger's and Vines's, give good sy-
nopses of the groups, as do also Atkinson's and Barnes's
works. For an account of the groups from the ecological and evolutionary standpoint, there is an admirable
little work, entitled "Evolution of Plants," by Campbell, which every school library should possess, and
which should be well read in connection with any
school course treating the natural history of plants.

A group of books which from some points of view
may be considered as text-books, but which I think
rather should be viewed as books of reference, are the
laboratory guides. These are books giving full laboratory directions for the practical working out of important topics, and the student is supposed to have them
open before him as he works. The great objection to
them as a class is that they necessarily presuppose
certain materials, and these it is by no means easy to
provide; and the restriction they impose upon a good
teacher is unbearable. On the other hand, as suggestions for the construction of guides by the teacher
for his own class, they have much value, and it is
chiefly for this purpose the guides in Part II of this
book are offered. Of the laboratory guides, one of
the most recent and excellent is Spalding's "Introduction to Botany," which, however, gives no attention
to practical physiology, though it has excellent summary
chapters containing much ecology. Another is Setch-

ell's "Laboratory Practice for Beginners in Botany,"
which is confined to the higher plants and excludes
practical physiology, though it gives great attention to
ecology.   It is prepared upon the unusual plan of
telling the student in detail what he will see — a plan
that few teachers consider pedagogically profitable.
MacBride's "Lessons in Botany" is excellent within its
limits, but it is too exclusively structural.   Older books
of this class, but excellent and influential nevertheless,
are the botanical part of Huxley and Martin's "Elementary Biology" and Arthur, Barnes, and Coulter's
"Manual of Plant Dissection."   The latter work, in
particular, has been much and profitably used in this
country.   Many practical laboratory directions are
given in several of the text-books to be mentioned
below, notably Bergen's, Barnes's, and Atkinson's.

Intermediate between these laboratory guides and
true text-books come books that are primarily guides
to observation, though they may also give much information, and to some extent are usable as text-books.
A very notable and excellent work of this sort is L. H.
Bailey's "Lessons with Plants," a book replete with
suggestions, new points of view, and valuable information.   Another in the same spirit, but less elaborate,
is Miss Newell's "Outlines of Lessons in Botany" with
its accompanying valuable Readers.

We pass next to consider text-books proper — works
intended to be studied fully and carefully by the indi-

vidual student. Such books were formerly all-impor-
tant and all-sufficient in botanical education. With
the rise of the laboratory method of instruction they
fell into disfavor, and many teachers attempted to teach
without them or with the aid only of laboratory guides,
on the ground that the student should learn only from
nature. Experience, however, is showing that laboratory study, while absolutely essential for the training of
natural powers and the correct understanding of natural
facts and phenomena, is, nevertheless, not alone sufficient; for, dealing as it necessarily does, even at its best,
with a few selected types, the view it gives of the subject is more or less disconnected and incomplete, the
more especially since many topics of the greatest importance cannot for practical reasons be introduced into
the laboratory at all. Of course, instruction by lectures,
or talks by the teacher, partly overcomes these draw-
backs; but I, in common with other teachers, have
found after trial of different plans that it is an immense
advantage to the students to have some good book to
which they can turn for additional information, and for
correcting the many errors and distortions that inev-
itably arise from lectures and laboratory work alone.
It is well to require students to read such a book with
great care. It does not matter whether or not the book
follows the same course as the teacher's instruction,
though the more nearly they correspond the better.
This book should be a text-book in the true sense, a

book of botanical texts, as clearly, attractively, inductively, synthetically written as possible. It should not be complicated by laboratory directions or pedagogic matter, all of which belong in a separate work. Indeed, the later text-books show a tendency to separate the text-book proper from the laboratory guides and directions, and to make the former simply an attractive and instructive reading book. Such a separation has always been shown in the German text-books, and is carried out in this country in Barnes's recent "Plant Life," and the logical extreme of this principle has produced the present work. In the use of the text-book there is one golden rule, *i.e.* never to require reading in it upon any laboratory topic until after that topic has been studied in the laboratory. The laboratory study not only allows of much more intelligent and appreciative reading upon the topics there taken up, but each topic thus studied forms a point of vantage from which excursions into the unknown may profitably be made. We all know how much more any account of a country or city means to us after we have been there, even though we have seen but a small part of it; and it is so with accounts of organisms.

Of good text-books for use in elementary courses there are several. The latest published in this country is Atkinson's "Elementary Botany," a concisely written, modern, finely illustrated work, from the physiological standpoint, with much attention to mor-

phology and ecology, and constant practical directions for laboratory study. Similar in these respects
is Barnes's "Plant Life," a book but a few months
older; but this is less a laboratory book, and more a
work for reading. More recent than either is Vines's
"Elementary Text-book of Botany," a thorough and
valuable work, whose plan, however, hardly fits our
methods of instruction in this country as closely as
do those of Barnes and of Atkinson. Somewhat the
same may be said of a text-book which is perhaps
in the abstract the best that has appeared in recent
years, the "Text-book of Botany," by Strasburger,
Noll, Schenck, and Schimper. This book is written
by four leading scientific experts, and is one of the
most reliable, readable, and best illustrated of botanical books, and, whether used as a text-book or not,
it is an invaluable reference work that no library
should be without. It may be bought in two parts,
of which the first is of most importance to the average teacher. Another recent American text-book is
that by Curtis, "An Elementary Text-book of Botany." Another work, particularly adapted to schools,
is Bergen's "Introduction to Botany," which combines both text-book and laboratory directions, and is
modern in matter and spirit. For schools that have
scanty equipment and cannot give abundant time to
a course, this work is particularly advantageous.
Gray's "Elements of Botany" is in its own field a

most excellent work and one that can hardly be superseded; but its standpoint is not modern, and its morphology is in places out of harmony with present opinions. Another excellent work that has had wide use is Bessey's "Botany," of which there is an abridged edition called "The Essentials of Botany." Clark's "Laboratory Manual of Botany" is a work prepared in the modern spirit, but it has not been favorably received by those competent to judge of its merits from a scientific point of view. The various laboratory guides by Spalding, Setchell, and others have already been referred to. There are, of course, yet other text-books, including several written in England, in Germany, and in France. So far as text-books for elementary work are concerned, it is more likely that one written in any particular country will be better adapted to methods of instruction in vogue there than would be the case with one written in another country. And it must also be remembered that, owing to the advancement of science, those books written by active scientific experts, and those that are newest, are likely to be the best.

There remains yet one other class of books to be considered, — manuals for use in classification. The classic work for northeastern North America is Gray's "Manual," sixth edition. The system of classification represented by it, however, is being generally abandoned, and a new edition is needed. A special edition with leather cover and thin paper is issued for use in the field. Covering about the same ground and embodying the newer classification, though also embodying a system of naming of the plants not yet widely accepted, is Britton and Brown's "Illustrated Flora," in three volumes. This gives a simple outline illustration for each species. A kind of work that is very much needed, and which is sure in the future to be prepared, is one that will be at the same time a synopsis of classification and of natural history, giving the habits and marked adaptations of each species. For the Southern States there is Chapman's "Flora," and for the Rocky Mountain region, Coulter's "Manual." For the Pacific slope there is not yet a compact manual comparable with the above-mentioned, though Greene's "Manual of the Botany of the Region of San Francisco Bay," and Howell's "Flora of Northwest America," partly cover that region. Of reference works for such studies, there are many, of which Gray's "Synoptical Flora of North America" is the most important. For the study of garden plants, Gray's "Field, Forest, and Garden Botany," revised by Bailey, is the only work. In this connection may be mentioned monographs of special groups, of which by far the greatest, and one of the most splendid works in every respect that have ever appeared, is Sargent's "Silva of North America," in twelve volumes, exhaustively describing every species of tree in North America. Of a similar character is Eaton's "Ferns of North America," which describes, though on a less elaborate scale, all of our ferns. Of books designed as guides to the study of particular groups, there are a few of real authority. Underwood's "Our Native Ferns and their Allies" is one of these, and of course there are many popular works of this character, whose consideration hardly belongs in the present work. If the teacher is interested in other groups, or wishes other information about botanical books, he should consult the Professor of Botany in the nearest large university.

BIBLIOGRAPHY

List of works referred to in Chapter VII and elsewhere in this book. The prices are from publishers' lists, and are usually subject to discount. They are for bound copies. Care should be taken to secure always the latest editions.

Arthur, J. C., Barnes, C. R., and Coulter, J. C. Handbook of Plant Dissection. New York. Henry Holt & Co. 1887. $1.20.

Arthur, J. C., and MacDougal, D. T. Living Plants and their Properties. New York. Baker & Taylor. 1898. $1.25.

Asa Gray Bulletin. Bi-monthly. Washington, D.C. 50 cents a year.

Atkinson, G. F.
   (1) Elementary Botany. New York. Henry Holt & Co. 1898. $1.25.
   (2) The Study of the Biology of Ferns by the Collodion Method. New York. The Macmillan Co. 1894. $2.00.

Bailey, L. H.
> (1) The Survival of the Unlike. New York. The Macmillan
>   Co. 1897. $2.00.
> (2) Lessons with Plants. New York. The Macmillan Co.
>   1898. $1.10.
> (3) Garden Making. New York. The Macmillan Co. 1898.
> $1.00.

Bailey, W. W.   Botanical Collector's Hand-book.   Salem, Mass.
    George A. Bates. 1881. $1.50. A new edition is promised.

Barnes, C. R.   Plant Life, considered with Special Reference to
    Form and Function. New York. Henry Holt & Co. 1898. $1.12.

Bates, H. M.   The Naturalist on the Amazons. London. John
    Murray. 1892. 18s.

Beal, W. J.   Seed Dispersal. Boston. Ginn & Co. 1899. 40 cents.

Belt, T.   The Naturalist in Nicaragua. London. E. Bumpus.
    7s. 6d.

Bergen, J. Y.   Elements of Botany. Boston. Ginn & Co. 1896
    and later. $1.20.

Bessey, C. E.
> (1) Botany for High Schools and Colleges. New York.
>   Henry Holt & Co. 1885 and later editions. $2.20.
> (2) The Essentials of Botany. New York. Henry Holt & Co.
>   1896. $1.08.

Botanical Gazette.   Monthly. Chicago. The University of Chicago.
    $4.00 a year.

Britton, N. L., and Brown, A.   An Illustrated Flora of the Northern
    United States, Canada, etc. New York. Charles Scribner's
    Sons. 3 vols. 1896–1898. $3.00 a volume.

Bulletin of the Torrey Botanical Club.   Monthly, Lancaster, Pa.
    $2.00 a year.

Campbell, D. H.
> (1) The Structure and Development of the Mosses and Ferns.
>   New York. The Macmillan Co. 1895. $4.50.
> (2) Lectures on the Evolution of Plants. New York. The
>   Macmillan Co. 1899. $1.25.

Chapman, A. W.   Flora of the Southern United States, etc.
    American Book Co. 1897. $3.60.

Clark, C. H.   A Laboratory Manual of Practical Botany. New
    York. American Book Co. 1898. 96 cents.

Coulter, J. M.   Manual of the Botany of the Rocky Mountain
    Region.   New York.   American Book Co.   1885. $1.62.
Curtis, C. C.   Text-book of General Botany.   New York.   Long-
    mans, Green, & Co.   1897. $3.00.
Darwin, Charles.
    (1) The Origin of Species.   1885.   Sixth edition. $2.00.
    (2) The Power of Movement in Plants. $2.00.
    (3) Insectivorous Plants. $2.00.
    (4) Movements and Habits of Climbing Plants. $1.25.
    (5) The Various Contrivances by which Orchids are fertilized
        by Insects. $1.75.
    (6) The Effects of Cross and Self Fertilization in the Vegetable
        Kingdom. $2.00.
    (7) Different Forms of Flowers on Plants of the same Species.
        $1.50.   All published by D. Appleton & Co.   New York.
Darwin, F., and Acton, E. H.   Practical Physiology of Plants.
    Cambridge.   Second edition.   1897.   4s. 6d.
De Bary, A.   Translated by Bower and Scott.   Comparative
    Anatomy of the Vegetative Organs of Phanerogams and Ferns.
    Oxford.   The Clarendon Press.   1884. $5.50.
Detmer, W.   Translated by Moor.   Practical Plant Physiology.
    New York.   The Macmillan Co.   1898. $3.00.
Eaton, D. C.   The Ferns of North America.   Boston.   Bradlee
    Whidden.   1893.   2 vols. $20.00 per volume.
Eliot, C. W.   Educational Reform.   New York.   The Century Co.
    1898. $2.00.
Engler & Prantl.   Die natuerlichen Pflanzenfamilien.   Leipzig.
    W. Engelmann.   Now appearing in parts.   12 volumes are
    complete.
Forbes, H. O.   Wanderings of a Naturalist in the Eastern Archi-
    pelago.   London.   Sampson, Low & Co.   1885.   21s.
Geddes, P.   Chapters in Modern Botany.   New York.   Charles
    Scribner's Sons.   1893. $1.25.
Goebel, K.
    (1) Translated by Garnsey and Balfour.   Outlines of Classi-
        fication and Special Morphology of Plants.   Oxford.
        Clarendon Press.   1887. $5.25.
    (2) Organographie der Pflanzen.   Jena.   Gustav Fischer.
        Now appearing in parts.

Goodale, G. L.   Physiological Botany.   Vol. II of Gray's Botanical Text-book.   New York.   American Book Co.   1885. $2.00.

Gray, Asa.

    (1) Scientific Writings.   Edited by C. S. Sargent.   Boston. Houghton, Mifflin & Co.   2 volumes.   1889. $6.00.

    (2) Structural Botany.   Part I of Gray's Botanical Text-book. New York.   American Book Co.   1880. $2.00.

    (3) Manual of the Botany of the Northern United States. Sixth edition.   New York.   American Book Co.   1890. $1.62.   Field edition, $2.00.

    (4) Elements of Botany.   New York.   American Book Co. 1887 and later editions.   94 cents.

    (5) Field, Forest, and Garden Botany.   Revised by L. H. Bailey. New York.   American Book Co.   1895. $1.44.

    (6) How Plants Behave.   New York.   American Book Co. 1875.   54 cents.

Greene, E. L.   Manual of the Botany of the Region of San Francisco Bay.   San Francisco.   Curry & Co.   1894. $2.00.

Haberlandt, G.

    (1) Eine botanische Tropenreise.   Leipzig.   W. Engelmann. 1893.   9.25 marks.

    (2) Physiologische Pflanzenanatomie.   Leipzig.   W. Engelmann.   1896.   18 marks.

Howell, T.   Flora of Northwest America.   Portland, Ore.   Published by the author.   Now appearing in parts.

Hudson, W. H.

    (1) The Naturalist in La Plata.   London.   Chapman & Hall. 1892.   16s.

    (2) Idle Days in Patagonia.   London.   Chapman & Hall. 1893.   14s.

Huxley, T. H.   Science and Education.   Vol. III of his Collected Works.   New York.   D. Appleton & Co.   1894. $1.25.

Huxley, T. H., and Martin, H. N.   A Course of Elementary Instruction in Practical Biology.   London.   The Macmillan Co.   1886. 10s. 6d.

Kerner von Marilaun, A.   Translated by Oliver.   The Natural History of Plants.   New York.   Henry Holt & Co.   4 volumes. 1894–1896. $15.00.

Knuth, K. Handbuch der Blütenbiologie. Leipzig. W. Engel-
   mann. 1898. 2 volumes. 28 marks.
Lubbock, Sir J. Flowers, Fruits, and Leaves. London. The
   Macmillan Co. 1884. 4s. 6d.
Ludwig, F. Lehrbuch der Biologie der Pflanzen. Stuttgart.
   F. Enke. 1895. 14 marks, unbound.
MacDougal, D. T. Experimental Plant Physiology. New York.
   Henry Holt & Co. 1895. $1.00.
Müller, H. Translated by Thompson. The Fertilisation of Flowers.
   London. The Macmillan Co. 1893. 21s.
Newell, Jane H. Outlines of Lessons in Botany. Parts I and II.
   with Readers, Parts I and II. Boston. Ginn & Co. 1892.
   Outlines, 90 cents each ; Readers, 70 cents each.
Pfeffer, W. Pflanzenphysiologie. I. Stoffwechsel. Leipzig.
   W. Engelmann. 1897. 23 marks. An English translation
   is being prepared.
Plant World, The. Monthly. Binghampton, N.Y. $1.00 a year.
Rhodora. Monthly. New England Botanical Club. Boston. $1.00
   a year.
Sachs, J.
   (1) Translated by Garnsey and Balfour. History of Botany.
      New York. The Macmillan Co. 1890. $2.50.
   (2) Lectures on the Physiology of Plants. Translated by
      Ward. Oxford. Clarendon Press. 1887. (Now out of
      print.)
Sargent, C. S. The Silva of North America. 12 volumes. Boston.
   Houghton, Mifflin & Co. 1892–1899. $25.00 a volume.
Sargent, F. L. Corn Plants. Boston. Houghton, Mifflin & Co.
   1899. 75 cents.
Schimper, A. F. W. Pflanzengeographie auf physiologischer
   Grundlage. Jena. Gustav Fischer. 1899. 27 marks, unbound.
Science. Weekly. New York. The Macmillan Co. $5.00 a year.
Setchell, W. A. Laboratory Practice for Beginners in Botany.
   New York. The Macmillan Co. 1897. 90 cents.
Sorauer, P. Translated by Weiss. A Popular Treatise on the
   Physiology of Plants. London. Longmans, Green, & Co.
   1895. 9s.
Spalding, V. M. Guide to the Study of Common Plants. Boston.
   D. C. Heath & Co. 1895. 90 cents.

Strasburger, E.

    (1) Das botanische Practicum. Jena. Gustav Fischer. 1897.
    22.50 marks.

    (2) Das kleine botanische Praktikum. Third edition. Jena.
    Gustav Fischer. 1897. 7 marks.

    The second edition of the latter is translated by Hill-
    house as Practical Botany. New York. The Macmillan
    Co. 1897. $2.50. New edition promised.

Strasburger, E., Noll. F., Schenck, H., Schimper, A. F. W. Trans-
lated by Porter. A Text-book of Botany. New York. The
Macmillan Co. 1898. $4.50. In four sections, and Sections
I and II may be had together without III and IV. $2.50.

Underwood, L. M. Our Native Ferns and their Allies. New
York. Henry Holt & Co. 1888. $1.25. New edition
promised.

Vines, S. H.

    (1) An Elementary Text-book of Botany. New York. The
    Macmillan Co. 1898. $2.25.

    (2) Lectures on the Physiology of Plants. Cambridge. Uni-
    versity Press. 1886. 21s.

Wallace, A. R. Malay Archipelago. London. The Macmillan Co.
1891. 6s.

Warming, E.

    (1) Translated by Potter. A Hand-book of Systematic Botany.
    New York. The Macmillan Co. 1895. $3.75.

    (2) Translated into German by Knoblauch. Lehrbuch der
    Oekologischen Pflanzengeographie. Berlin. Geb. Born-
    traeger. 1896. 8 marks.

Wiesner, J. Biologie der Pflanzen. Vienna. 1889. A new edition is in preparation.

VIII. ON SOME COMMON ERRORS PREJU-DICIAL TO GOOD BOTANICAL TEACHING

ONE of the chief obstacles to the advancement of knowledge is the difficulty of securing the introduction of the results of new researches into general circulation. Errors once in possession of the field, especially if backed by the authority of some great name, persist long after they are disproven, particularly when easier to understand, or pleasanter to believe than the newer truths. I shall here point out some of the more prevalent errors in Botany, not including cases still in doubt, but only those on which competent authorities agree.

Very widely spread is one popular error about Botany; namely, that it is synonymous with the study of flowers, and hence of no great value except as an accomplishment of fashionable boarding-schools or an appropriate hobby for elderly persons of leisure. This belief is a natural one, for until lately it *has* consisted in this country largely in the study of flowers, and still does to far too great an extent. We cannot expect the error to be corrected until botanical courses generally represent the real condition of the science.

Another error, prevalent even among college teachers, is, that Botany cannot be taught as a science in the high schools because high school students are not mature enough, and cannot think. It is true that many of them do not think, but this is because their thinking powers are aborted by disuse or crushed to earth by the weight of incessant memory work. But experience shows that, given a fair chance, high school students can think, and are fully able to profit by even the most scientific treatment of the subject.

These, however, are but minor errors, though it is well for the teacher to be on the watch for them, and to attack them whenever they appear. More serious are the errors of botanical fact and theory current among teachers themselves. Thus, morphology, as commonly taught in our schools, is dominated by a rigid formalism, an inheritance from the idealistic system of Goethe, which implies modification within the limits of some distinct plan rather than modification in adaptation to conditions as they exist. It is commonly taught that the higher plant consists of root, stem, and leaf (with, perhaps, also " plant hair "), and that every part of it is composed of some one or more of these, which, like the chemical elements, may be variously combined and united, but retain their identity through it all. This view is a natural one where evidence is drawn from comparative anatomy alone, and most of those who have held it have been students in that line and not in embryology. It is a very poor working hypothesis, for it leads always to a blank wall blocking further prog-

ress, and to inconsistencies requiring the most artful
dodging.   The central point in this doctrine is the
belief in the comparative immutability of the nature of
the plant members or elements.   All modern research,
however, is denying this belief and replacing it by the
opposite principle, viz. that difference of degree of development passes over into difference of kind of structure, thus leading to the formation of new elements
or members which become centres of variation, modification, adaptation upon their own account, and more or
less independently of their original nature.   Thus the
ovary is composed of carpels, which originally were
spore-bearing leaves.   Now, when an ovary must vary
adaptively to some new influence, it does not need to go
back to consult the rules governing its behavior when it
was a set of sporophylls, but it responds as a unit, as
an ovary; it has itself become a member or element.
It is always necessary in morphology to keep plain the
difference between historical origin and present nature.
Historically, an American is an Englishman, but he
does not on that account now act or think as an Englishman; he has a new character, he is an American.
So, historically, the ovary is a set of leaves; but it does
not act like a set of leaves, but like what it now is, an
ovary.   The placenta is another good example of this.
Generally it is said to represent the united edges
of infolded carpellary leaves, but one has to perform
complicated mental gymnastics to interpret all placentæ

L

in this way.   The real explanation doubtless is, that,
although the placenta did thus originate, it has attained
to independent dignity as a morphological element, and
proceeds to act as a unit, varying and adapting itself
to its conditions, largely independently of its original
nature.   Of course there are all degrees of this independence of original nature ; and while some structures
have broken away entirely from it, others are more or
less bound by it ; but the recognition of the principle,
really the fundamental principle of modern morphology, is very important.   Formalism in morphology,
based upon comparative anatomy, must be modified by
realism based upon embryology.   Unfortunately there
is as yet no authoritative work in English treating fully
this subject.

Perhaps the greatest of all current morphological
errors is that which attempts to homologize ovules and
pollen with something on the green shoot.   It is often
taught that the ovule represents an altered piece of the
edge, perhaps a tooth, of the leaf, while the pollen is the
parenchyma with epidermis rounded out to form the
anther.   It is, however, now known beyond any doubt
that the ovule (strictly, its nucellus) is a spore-case, a
lineal descendant of the spore-cases of liverworts, and
hence is much older than the differentiation of the parts
of the leafy shoot ; and the same is true of the pollen and
anther, which represent also ancient spores and spore-
cases.   Hence the pollen and ovules are not modified

parts of a vegetative shoot, but independent structures of long ancestry and independent dignity.

Another common error is that of attempting to homologize the parts of the stamen and of the carpel with the parts of the green leaf. Thus, it is often said that the filament represents the petiole, and the anther the blade; while in the carpels the ovary wall is said to be the blade, the style the elongated tip, and the stigma the extreme end turned back; and much mental ingenuity is needed to show how some of the more specialized styles and stigmas fit into this scheme. In fact, while the precise origin of the stamens and carpels is not beyond doubt, this much is certain, that they are the descendants of the sporophylls of cryptogamic plants, and there is a very strong probability that these sporophylls never have been green foliage leaves; and even if they were, it was at a time before the differentiation of the modern specialized foliage leaf with its blade, petiole, and stipules. It is, therefore, correct to regard carpels and stamens as morphologically leaves; but they must be viewed as having followed a course independent of that of foliage leaves, each developing the parts necessary to its function without any regard to the developments of the other. Hence it is not possible to homologize the parts of one with the parts of the other.

Another very common error, perpetuated by its use in all systematic works, is, that inferior ovaries represent united carpels and calyx. Nothing could be more thoroughly disproved, or is more easy to disprove by embryological study than this. Embryology shows that, in the great majority of cases at least, the inferior ovary is simply a receptacle which has grown up into a cup, carrying all the other parts upon its top, the carpels coming finally to form simply a roof over the cavity of the ovary (as shown in Fig. 28). This fact at once disposes of many of the inconsistencies inseparable from the "calyx-adnate" theory. Again, where a calyx- or corolla- or perianth-tube is formed, it is usual to consider that this tube consists of united sepals, petals, etc., but it is probable that only the free parts or teeth of the corolla or calyx represent the original distinct petals or sepals, while the tube is a band of leaf tissue that grows up as a ring leaf, bearing the separate leaves on its top; it is thus a new structure and not the united bases of the old perianth parts.

There are also some minor errors current, of which the following are most important: First, it is usually supposed that root, stem, and leaf of the higher plant are members of equivalent worth. In fact, this is not the case, for root is in every way much more distinct from stem and leaf than these are from one another. The best division then is into root and shoot, with the latter differentiating into leaf and stem. Another error is, that the higher plant is made up of certain elemental parts called phytomera, each of which is composed of a joint of stem and one or more leaves. The support for this idea is found partly in the jointed appearance of many plants like grasses, and partly in the fact that the so-called phytomer is usually the smallest part of a plant that will grow. The latter, however, is a purely physiological phenomenon of no morphological signifi- cance; a piece of stem can usually put out roots, and some leaf surface is necessary to make food to enable the plant to continue to grow. The jointed appearance is purely incidental; the nodes are the places where the fibrovascular bundles branch to run out into the leaves and to unite with one another, and hence the node and its accompanying internode have simply an anatomical and not a morphological meaning. Embryology shows that the plant, so far from being made by a series of phytomera growing one out of another, is made by continuously growing vegetative points, throwing off laterally certain superficial portions which become leaves, in whose axils the points branch.

The true morphological relationships of the parts of the higher plants are expressed in the following table :—

Root

Shoot
{
  Stem . . . .

  Leaf
    Foliage { Blade / Petiole / Stipules

    Floral { Petals and Sepals . . . .

    Sporophylls { Macrosporophylls (Carpels). / Microsporophylls (Stamens).

Sporangia (containing Spores) { Macrosporangia (Ovules) . . / Microsporangia (Anthers). . }

The Flower

Among errors very dear to us is the belief that
monstrosities are reversions to an earlier condition, and
hence good guides to the past history of organs or
species. It is true they may be, and of course often
are; but they so frequently are not that great caution
must be exercised in using them as guides to phylogeny.
If the turning of a petal green is taken to prove that
the petal was once a foliage leaf, then the turning red
or yellow of the leaf under the flower of a tulip must be
taken to prove that this leaf was once a petal, which is,
of course, not to be believed; hence the turning green of
petals means nothing more than a disturbance of nutrition conditions. This principle applies to the cases
where carpels become leaves and the ovules leaf-like
bodies, which need not mean that these were once of a
green leaf nature, but only that the plant has for some
reason unknown made its materials build green leaf
tissue instead of carpellary tissue at that place.

Very common and serious are physiological errors, of
which perhaps the most widespread is the belief that
animals and plants are the exact opposites of each
another with reference to the taking in and giving off
of the two very important gases, carbon dioxide and
oxygen. In a general way this is true, but not in the
sense in which it is usually meant. In fact, in all of
their processes of growth, movements, etc., animals and
plants behave precisely alike with reference to these two
gases, in both cases taking in oxygen and giving out

carbon dioxide. But it happens that green plants have an additional power, utterly lacking in animals, to form their food from certain gases, minerals, and water, and in this process (photosynthesis, or assimilation) carbon dioxide is absorbed and oxygen is given off. In green plants, in bright light, this process is so very much more active than the process of respiration, that the plant as a whole does give off much more oxygen than carbon dioxide, but in darkness the food-making stops, and the plant then gives off carbon dioxide precisely as animals do. It is therefore only in virtue of their possession of this single extra power that plants reverse the process of animals; in nearly all others of their important vital actions they behave like them.

Much misunderstood is the nature of plant food. It is generally taught that plant food consists of carbon dioxide, minerals, and water. If by food one means anything taken into the organism, this is correct; but if by that term one means the substance out of which the organism builds up new tissues, repairs waste, and obtains energy for its own vital work, then this is incorrect. In reality the plant has the power, lacking in animals, of absorbing the carbon dioxide, water, and minerals, and of making from these starch or a related substance; and this starch is then used as food in essentially the same manner as animals use it. It may be said, then, that plants form their food from the raw materials, which properly are not food at all. Of course animals are entirely dependent for their food upon that made by plants.

Another error is the assumption that the carrying of pollen from flower to flower by insects is a part of the process of reproduction, and this is intensified by the common expression of "cross-fertilization" used to describe the process. But really this process has nothing directly to do with the act of fertilization, but it is simply one of the methods the plant has adopted to overcome the difficulties imposed by the sessile habit; that is, it is the mode of locomotion of the male to the female element, and is much better described as pollen-locomotion or cross-pollination.

When one studies the phenomena of irritability, he usually passes through a stage in which he believes that plants possess a certain intelligence. The more careful study of the phenomena, however, leads to the conclusion that it is not intelligence they possess, though they have a power producing some apparently similar results. Irritability is more nearly comparable with reflex action, and even with instinct in animals, than with their consciousness and intelligence. It may be said that out of one and the same property in the original protoplasm, animals have differentiated reflex action, instinct, and intelligence, while plants have developed irritability.

It is only by keeping in touch with the most modern and authoritative books that the teacher can correct the older errors.

PART II

AN OUTLINE FOR A SYNTHETIC ELEMEN-
TARY COURSE IN THE SCIENCE
OF BOTANY

INTRODUCTION TO PART II

THE principles that have controlled the construction of these outlines have been set forth fully in the preceding chapters, and in synopsis are as follows: The ideal is to guide the student to the optimum return of sound scientific training and thorough botanical knowledge for the time and strength he can put into the work. They are hence a study in educational economy, with three principal phases: first, the selection of the most vital and illuminating topics; second, a synthetic treatment of the science, with topics arranged in such an order as to throw most light upon one another; third, the presentation of the topics in such a form as to draw out the student's faculties the most quickly and thoroughly. They have not been worked out regardless of practical considerations, but with constant account of them, and with special effort to show how the restraints imposed by them may be minimized.

The general plan of the entire course is the double one used by many teachers: a first division treats of the principles of anatomy, morphology, physiology, ecology; and a second, of the structure and adaptations of the principal groups of plants from the lowest Algæ to the higher Phanerogams.

In Division I a beginning is made with large, simple, somewhat familiar objects, requiring no tools, but only the undivided attention of eye and thought. It is sought first to form the scientific instinct, — the habit of observation, comparison, and experiment. Later, the simpler tools are gradually introduced, and the less familiar materials and topics. Experiments, arranged to be tried with apparatus as simple and inexpensive as possible, are introduced along with the particular structures they throw most light upon. Every new topic is presented to the student in the form of a problem so arranged as to be solved through proper inductive processes by his own efforts. Practically, they form a series of original investigations. These problems are introduced by questions asked in a form to direct attention to the leading facts and phases of the subject. Indeed, the form of the questions is one of the most important features of such outlines as these, for they may be made to dissipate or to conserve energy, and are the chief means at command of the teacher for directing observation and comparison along the most useful lines. It is by no means only the easiest or most familiar topics and experiments which are here recommended, but a direct attack has been made upon the most fundamental and important.

Since it is of the utmost importance to a proper
conception of the meaning of the modern science
that the student's introduction to it should be through
the study of plants alive and at work, and since, in
our climate and especially in city schools, much accurate field work is impracticable, the tracing of some
living plant through its life cycle forms the best beginning known to me. Since plants develop from
the seed with relative rapidity, and the phenomena
of their growth, movements, etc., can be readily seen
and experimented upon, the germination of the seed
affords the best starting-point. After a single plant
is thus followed through its cycle from seed to seed,
the modifications of this typical form in response to
different habits are taken up, and then the different
members — leaf, stem, root, flower, fruit — are studied
in detail as to functions, structure, and ecological
modifications. Practically, most general botanical
principles may be worked out best in the higher
plants, because these are larger, more familiar, and
easier to obtain.

In Division II, living plants which may be studied
alive, and even seen in their native haunts, with
attention called to their habits, are used in almost
every instance. With the knowledge and training ac-
quired in Division I, the students work through this
second division with great profit, and it is by no
means inferior in value to the former. Here the

lower, or cryptogamic, plants receive their proper attention, and here, too, is the proper place of classification.

In using these outlines, it is by no means expected
that any teacher will try to follow them exactly;
although at the same time, in view of the amount of
care, based upon much trial and experiment, which
has brought them into their present form, one should
have good reasons for the changes he makes. Of
course many practical considerations are likely to
make it impossible to provide the exact materials
called for, or to take up the topics in precisely this
order. Indeed, it is in general very hard to provide
the materials to fit any particular set of outlines, and
it is much easier and more logical to make outlines
to fit the materials. These outlines are rather a series of suggestions, based on considerable experience,
representing useful selection and treatment of topics and expression of problems. They may serve
as a basis or as models for the teacher in the construction of new outlines of his own, differing from
these little or much as he pleases. Certainly, I think,
a special outline should be drawn up by the teacher
each week to fit his particular mode of teaching,
the material available, the state of advancement of
his class, etc., and a copy of this should be placed
before each student who is to be held responsible
for the complete working out of all that is called for

upon it. Directions in some form or other must be given the students by the teacher; when spoken, some students do not hear them, others forget them, but the written outline keeps them before all. So great is the advantage of these weekly guides in economizing the teacher's time and strength, and in giving definiteness and direction to the student's work, that there is in my experience no pedagogic device of greater worth. There is not the slightest objection to them on the score of weakening the student's self-reliance, and when given a proper form they become a great stimulus to him. They completely deliver the teacher from that otherwise familiar but awful question, "What do you want me to do next?"

The experiments here given are such as seem to me indispensable. Experiments much easier to try are given in various books, but many of them are on comparatively unimportant topics; and it is worth while to take some extra trouble to illustrate subjects so fundamental as those here recommended.

The entire course as given in the outlines has been carefully adjusted as to time, and is worked out by my own students in a college year, with four, or somewhat more, hours in the laboratory, one demonstration hour, and one lecture a week. If but half, or much less, of this time can be given, the teacher will naturally select from the list the more important topics.

DIVISION I

THE PRINCIPLES OF THE SCIENCE OF BOTANY

I. The Anatomy of the Seed

1. *a.* Study the outside of the dry Lima Beans; compare several specimens, and observe what features are common to all and what are individual; minutely observe : —

   (1) What is the typical shape?
   (2) What is the color?
   (3) What markings have they?

   Answer, as far as possible, by drawings made twice the natural size; add notes to describe features which drawing cannot express.

   b. Remove the coatings from soaked seeds.

   (1) What effect has the soaking had upon the markings, size, and shape?
   (2) How many coats are there?
   (3) Do the external markings bear any relation to the structures inside?
   (4) What shapes have the structures inside, and how are they connected with one another?

Answer as before by drawings and notes, the
former natural size.

2. Study fully in the same way the Horse Bean.

3. In a concise paragraph describe the resemblances
    and the differences of the Lima and the
    Horse Beans.

Materials. — White Lima Beans (*Phaseolus lunatus*) and
Horse Beans (*Vicia faba equina*), about six to a student, may
be bought in all large seed stores ; half should be soaked over
night. Windsor Beans may be used in place of the Horse
Beans, and other kinds will do ; but those selected should be
large, and such that in one the cotyledons come above ground
in germination, and in the other they remain below.

Pedagogics. — This outline can be completed by the average
student in two two-hour periods, but three are much better ; if
necessary, Exercise 3 can be completed at home from their
drawings and notes, but it is better worked out with the seeds
in hand. No tools except a pocket knife are needed, not even
a lens. In a large laboratory division general guidance to the
whole class, as well as individual help, not too much of either
at first, should be given, as recommended in Chapter III. The
exercises in this outline are principally to teach beginners : —

(I) To see a natural object as it is, correctly and completely.

(II) Through comparison to eliminate accidental and individual features, and thus to distinguish essential from unessential characters.

(III) To represent clearly to another what is seen, for this
purpose using words or drawings according as the one or the
other is the more expressive.

(IV) A knowledge of the anatomy of some typical seeds.

Following are points of importance in the teaching of these four essentials : —

(1) *On Observation.* — It is of first importance that the student learn to see natural facts absolutely as such, uninfluenced by any explanation of them. Hence he should be kept at work upon the Lima Beans until he has clearly seen (as shown by his drawings and notes, and under questioning) which of his specimens are average or most typical ; what their shape and color is ; the radiating markings, stopping short of the edge ; faint concentric markings (not always visible) ; on the concave edge a large scar, at one end of which is a tiny pit, and at the other a tiny raised yellowish triangle, which continues into a faint ridge ending in a more raised portion, the latter making an angle as seen from the side. Observation consists not only in seeing all these things, but in seeing them in their proper relative positions and connections. Names and uses should not be given until after the things have been seen, some curiosity aroused as to their use, and a need felt for names for them.

On removing the seed-coat, the student should see that this is single (actually two united, though usually he cannot see that) ; also the thick line representing the ridge he saw outside ; and the lack of any connection between exterior markings and the structures inside, excepting only the position of the micropyle over the end of the hypocotyl (not of course at first using those terms).

In the embryo he should see that hypocotyl and epicotyl are united, one the continuation of the other ; that the cotyledons are lateral growths of the hypocotyl ; and that the plumule consists of a short stalk bearing two folded veined leaves, one partially enclosed in the other.

(2) *On Comparison.* — The student should see that some

cracks and folds are simply due to individual differences in the mode of drying, etc., and that shape and size are variable, though within limits. In his treatment of Exercise 3 he should be led to distinguish clearly resemblances and differences, and to describe them separately.

(3) *On Representation.* — The general principles of this part of the work are discussed in Chapter IV. Observation should be fully made before recording is begun. As to drawing, the students should first be allowed to do the best they can unaided, judging for themselves how many and what kind of drawings are necessary to show completely the seed and its parts. They should be made to complete a subject the best they can before it is examined or criticised by the teacher; this is to inculcate self-reliance. After they have done their very best, their work should at once be examined, and wherever it shows marked deficiencies they should be encouraged to look and try again. After they have finally done all they can, the teacher should, step by step, carefully explaining the logic of each point, show them the best way he knows for representing the objects, with which they may compare their own efforts; this may well be done for them all together on a blackboard; they are, after their own trials, in a position to profit by all of the advice thus given. A good representation of the Lima Bean as an example for beginners is shown in Fig. 12, though the faint radiating lines might have been added. But while representation is made thus important, the teacher must not go so far as to make a fetich of it; for after all it is but a means to an end. At first only clear diagrams should be insisted upon — shading, etc., may better come later. It is, moreover, very important not to insist upon too many things at once, as this tends but to confusion; earlier exercises may well be left somewhat incomplete for this reason. The descrip-

tions in words should be studies in clearness and conciseness ;
but perfection cannot be expected at the start. From the first,
rough sketches should be forbidden. Few drawings may be

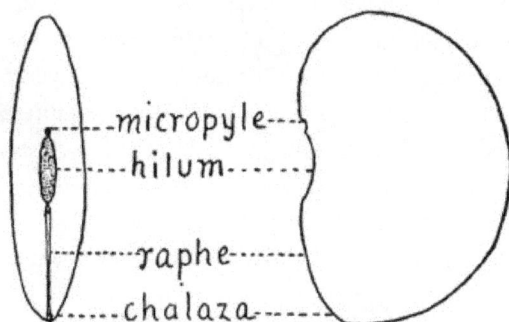

FIG. 12. — Good drawing, by a beginner, of Lima Bean, ×1¼.

made, but in these every line and spot should have its meaning, and nothing admitted for which there is not an equivalent
in the seed. Outlines should be firm, clear, and complete, and
haziness not permitted. The drawings should not be a composite made up from several specimens, but an accurate drawing of a typical specimen.

(4) *On gaining Knowledge of Seed Anatomy.* — Their observation gives them the lead-
ing facts of seed structure.
After they have seen and
represented the structures,
the teacher should lead
them (using the blackboard
while showing them better

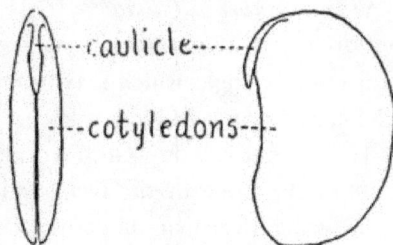

ways of drawing) to ask
what is the use or other
FIG. 13. — Good drawing, by a beginner, of
embryo of Lima Bean. Actual size.
meaning of each part ; and as they have no data for determining any of these, except, perhaps, the hilum, the early

life and development of the seed must be briefly described to
them with reference to the use of each part. This applies in
particular to the markings; the use of parts of the embryo
they will learn for themselves later. Along with this, and after
making them feel the need for single terms to describe the
different features, the proper names may be given them for
the parts, and these terms may be the better impressed upon
them if accompanied by side remarks upon their etymology,
etc. Of course names and uses should be carefully recorded.
Terms needed are *Coats*, *Hilum*, *Micropyle* (*Strophiole*, very
small in bean), *Raphe*, *Chalaza*, *Embryo*, *Cotyledons*, *Hypo-*

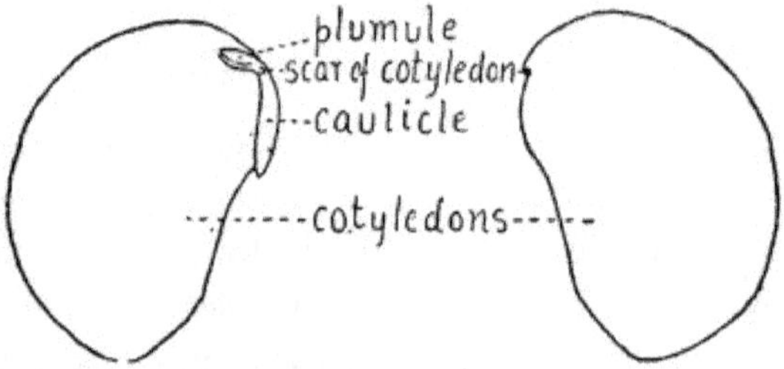

FIG. 14. — Good drawing, by a beginner, of embryo of Lima Bean laid open.
Actual size.

cotyl (or *Caulicle*), *Plumule*. (Of course the chalaza itself does
not appear in the seed, but its position is shown by a slight
projection or angle, which may be called the " chalazal angle.")
The names may be added as shown in Figs. 12, 13, 14.

The teacher will do well to work up for himself the development of the ovule in the Bean, which can very easily be done
from String Beans of different sizes. By a series of outline
diagrams he can then make clear to the class the exact meaning of the peculiarities of form and markings in the seed. This
would form an excellent topic for investigation by some of the
brightest pupils.

II. The Anatomy and Morphology of the Seed

4. *a.* Study the outside of the Horse-chestnut (a typical specimen), and minutely observe : —

> What is its shape, its color, and its markings?

> Does it show any feature not in the Beans?

Answer, as before, by drawings and notes. Select for yourself the best scale.

   b. Remove the coatings from a soaked seed, and observe : —

> (1) How many coats are there, and how are the markings related to structures inside?

> (2) What shapes have the internal parts, and how are they connected with one another?

> (3) Are there any new parts or features not present in those already studied?

Answer, as before, by drawings and notes. Carefully separate with the fingers all parts that can be forced apart without tearing.

5. In a similar way study the seed of the Morning-glory.

6. In a similar way study the grains of Corn.

> Where parts cling too closely to be separated by the fingers, use a knife, and try clean median sections.

7. *a*. In a concise and tabular form, compare as to the
chief resemblances and differences the four
kinds of seed you have studied, — the Bean,
Horse-chestnut, Morning-glory, and Corn.

   b. Construct a series of four diagrams, showing by
corresponding colors the relative development
of the equivalent parts in the four embryos.

   (Place this series on the upper half of one page,
and leave the remainder for a related series
to come later.)

Materials. — The Horse-chestnuts should be soaked for a
week ; if then the cotyledons do not separate readily, immersion in hot (not boiling) water for a few minutes will make
them. For the purposes of this exercise it is a most valuable
seed, and every effort should be made to obtain it. Morning-
glory seeds, the largest size, which may be bought cheaply by
the ounce in seed stores, should be soaked only four hours.
Though this seed is small, it is hard to find a larger one which
is as instructive. The Castor Bean (*Ricinus*) germinates badly,
and hence cannot be followed into its later stages, while Four-
o-clock is puzzling through presence of the fruit. A lens will
make the Morning-glory sufficiently clear. Corn, soaked over
night, must be studied chiefly by sections.

Pedagogics. — This outline will require at least three two-
hour periods, with some outside work.

   Its object is to continue training in observation, and to form
an introduction to morphology. As to observation, after their
previous experience, the students will readily find in the Horse-
chestnut everything on the seed-coats, including the fibro-vascular bundles on the hilum. The coats are two united, but

will seem to them as one. They should see that the hypocotyl does not lie against the cotyledons as in the Bean, but is separated from them in part by a seeming pocket of the coats (really due to a folding of the young ovule enclosing part of the coats) ; and that the seeming hypocotyl really splits down part of its length and has the plumule at the bottom of the split. In the Morning-glory they should find (with help of a lens) micropyle and raphe as well as hilum, and the jelly-like endosperm and the two cotyledons. In the Corn they should see, in addition to the other parts, the remnant of the silk (style) and the leaves of the plumule, on a failure to see which they should be reminded, not simply to look at things, but also to move and separate them.

Of the utmost importance in biology is morphology. Practically, it consists chiefly in recognizing the original nature of parts, no matter how much disguised by changes of size and shape. Its best index is the relative positions of parts. The Horse-chestnut is good to begin with, for the student may be made to work out for himself, by careful comparison with the construction of the embryo in the Bean, that what he at first always takes for "hypocotyl hollowed out with the plumule at the bottom," is really largely stalks of the cotyledons, while the hypocotyl is only the part below the plumule. In the Morning-glory he is apt at first to mistake the very leafy cotyledons for plumule, but can be led to work out their true nature. And in the Corn he can thus discover that the shield-like body is cotyledon. (Actually there is some slight doubt on this point among experts, but it is probably true, and can be so treated, with a caution to the students.) The observation of the remnant of the style on the corn grain, and their inability to find any equivalent for it on the other seeds, may be used to introduce an explanation of the composition of this

grain as ovary united to seed; and they may be led to notice
that the micropyle is not present, and that the scar of attachment, while functionally a hilum, is not strictly so morphologically. The occurrence of food substance outside of the
embryo in Morning-glory and Corn should be used to make
them seek for it in Beans and Horse-chestnut, and thus to
work out the differences between "albuminous" and "ex-
albuminous" seeds.

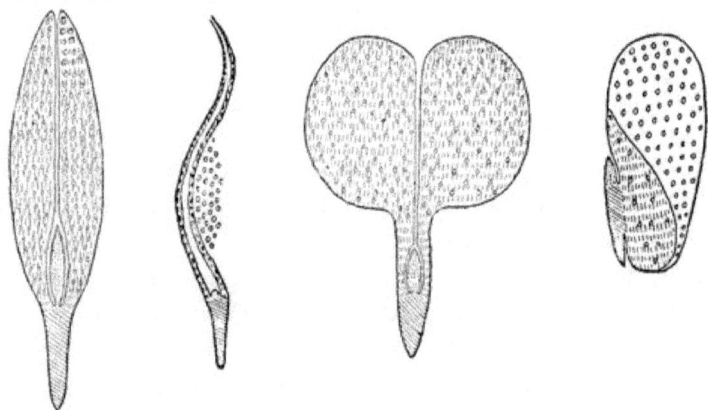

FIG. 15. — Diagrammatic figures of embryos of Lima Bean, Morning-glory,
Horse-chestnut, and Corn. shaded to show morphologically equivalent parts.
Diagonal lines = hypocotyl; vertical lines = cotyledons; dots = plumule;
circles = food substance.

Of the greatest morphological value is the Exercise 7 (b),
which is one of the best I have ever tried for inculcating a
true idea of morphology, the more especially when combined
with similar diagrams of germinated stages of the same seeds.
In making these diagrams all unessentials should be omitted,
and an effort made to represent only the principal corresponding parts placed in corresponding positions. The diagrams
should be somewhat as illustrated in Fig. 15, except that the
equivalent parts can be brought out much better by colored

crayons than by the black and white lines here made necessary by the method of engraving. The food substance may
be represented by small circles of blue, or of some other color.

In 7 (*a*) they should not run to details of little importance,
and resemblances should be emphasized as well as differences.

About this time a tendency will manifest itself to turn the
precious laboratory hours into a drawing lesson ; this must be
firmly met by making it plain that the laboratory time is for
observation and essentials of recording, and that all niceties
must be added in outside time, though rapid workers may
naturally be permitted some liberty in this respect.

No new terms are needed except *endosperm* and *albumen*,
the latter only in connection with the compounds "*albuminous*" and "*ex-albuminous*." It is best not to give at all any
terms of very limited application, such as "scutellum." As to
tools, a knife or scalpel may be used for sectioning, and a lens
for the Morning-glory (after they have tried to work without
it). Tools and terms should be given only after students have
been made to feel the need for them.

While the subject of the structure of the seed is fresh in
memory it will be well for them to read the very fine chapter
on this subject in either of Dr. Gray's text-books. Many
additional exercises on seeds are outlined in Spalding's and in
Setchell's books ; and if other materials for the next following
exercises are wanting, or some students manifest a special
interest in the subject, these may well be introduced. But for
most students it is more profitable to pass on to other subjects
than to spend additional time upon this. Considerable simple
physiological experimentation upon the growth of seeds in
relation to temperature, light, moisture, and oxygen, is possible,
and described in Bergen's Botany. Most of these facts thus
proved, however, are not specially characteristic of seeds, but

apply to other stages of growth as well.   A study of the storage
of nourishment in seeds is important, a subject well treated in
Bergen.

III. The Locomotion of Seeds

8. The seed is the locomotive stage of the plant.
Seeds have no power of independent movement,
and hence can secure locomotion only through
being carried by some of the moving agencies of
nature.   To fit them thus to be carried, adaptively constructed appendages have been developed.

  a. What are the different moving agencies of
    nature which can carry seeds?

  b. Study the ten seeds supplied to you.   In
    each case find out and record : —

    (1) What part produces the special
        appendages?

    (2) To what moving agency are the
        appendages probably adapted?

    (3) What accessory features of shape,
        weight, etc., to aid the appen-
        dages, are found in the seed
        itself?

  Make only outline sketches fully labelled.

9. Write a concise essay (of not more than two hundred and fifty words) upon the principles, deduced
from your laboratory work, from the lectures, and

from your reading, of the anatomy, morphology, and ecology of the seed. This is to be handed in (here the date).

Essays are to be written in ink in the Essay Book only. Each is to be preceded by a tabular outline of its contents.

Materials. — These must chiefly be collected beforehand in the summer. Good ones are Maple, Asclepias, Agrimony, Spruce or Pine, Desmodium, Ptelea, Elm, Xanthium, Burdock, Bidens, Dandelion, Tecoma or Catalpa, Galium, Castor Bean, Geranium maculatum. It is desirable to have, as in this list, some seeds and some "fruits." The museum collection, including some of the more remarkable kinds, will here be very valuable. The use of berries and other pulpy fruits should be explained by the teacher, since it could hardly be imagined from laboratory study; their morphology more properly comes later in the section " Fruits."

Pedagogics. — This exercise is for further training in observation, comparison (morphology), and for an introduction to ecology (*i.e.* adaptation to conditions of the external world).

In morphology the student should trace out from exactly what part the appendage is developed, whether from seed-coat, ovary, style, or calyx. To aid in this, the teacher must give some account of the structure of the flower and fruit. To distinguish whether a given structure is seed-coat or ovary, dissection will be necessary. He will thus discover that what are ecologically the same structures may have very different morphological origins ; from which he should be led, after the ecological use of the parts has been learned, to infer the great importance of function in developing structures.

In ecology, since the study is in the laboratory, and not out of doors (as it would much better be), the students can do little better than guess at the use of the different appendages. They can, however, be much helped by recalling facts already known by observation, as to the carrying of maple, willow, and other seeds and fruits by wind, and the sticking of seeds to their clothes in their walks through pastures, and also by some simple experiments, suggested by the teacher, upon the different seeds in the laboratory. This work will give them an introduction to theorizing — a habit of the greatest value in biology if kept checked by rigid observation or other confirmation, and of the greatest disaster if allowed to become merely untested guessing. In this case, since confirmation from outdoor observation is impracticable, the correctness of their theories will need to be tested by reference to the teacher, who should be thoroughly informed upon the subject; but it should be made plain that the teacher's knowledge is not better than their own observation, but only a substitute enforced by circumstances.

In Exercise 8 (*a*) the students will think of *wind, animals*, and probably *water currents*, to which hints from the teacher may cause them to add *throwing by spring-apparatus*, which include all of importance.

In their drawings the important locomotive appendages should be clearly brought out; for example, in the Burdock, half of the students will not represent the hooked tips, though they are plainly visible; in such cases they should individually be told they have missed something important, and left to seek until they have found and correctly represented it.

A fully illustrated account of this very important subject of seed locomotion, one of the most interesting of all botanical topics to most people, should be given in a talk or lecture.

Books relating to it may be found cited in Chapter VII. Other adaptations in seeds may also be taken up, such as their protection against animals until ripe ; how they absorb water ; how some seeds plant themselves, etc.

For the essay on the seed, consult the advice in Chapter IV. It is convenient to have a special book for the essays, uniform with the laboratory book. After they have done their best on this essay, it is well to read them one written by the teacher as a model. Following is one I have read for this purpose to my students : —

THE SEED

General Function.

Structure, — Coats, Embryo, Endosperm.

Locomotion.

The seed is a portion of plant substance specialized for reproduction and locomotion. Under a great variety of forms, sizes, and colors, seeds have in common the coats, embryo, and endosperm. The coats, one or two, are protective, and the outer usually shows the scar of attachment to the pod (hilum), a pit by which the fertilizing pollen tube entered (micropyle), and a ridge through which the nourishment was distributed (raphe and chalaza). The embryo is the young plant, and consists of stem (hypocotyl), on which are placed laterally one or two leaves (cotyledons), and which merges upwards into the bud (plumule). The endosperm may be stored in the cotyledons, making them thick, or around them, or in both ways.

Locomotion is as essential to plants as to animals, and since the adults cannot move, the seed is generally used as the locomotive stage, and to it appendages are added to cause it to be carried by some of the natural moving agencies. These appendages may be outgrowths of the seed-coat, or of ovary, style, or calyx, retained for the purpose. They may consist of wings or plumes to utilize the wind, hooks for attachment to the fur of animals, pulp to be eaten by animals, or may be absent altogether, in which case the seeds are often projected by the springing of elastic tissues.

IV. The Germination of the Seed and Growth of the Embryo

10. Study the germinating Lima Beans, and, in comparison with your records of the ungerminated bean, observe : —

   (1) Whether all seeds have developed at the same rate. If not, why not?
   (2) Where and by what force has the coat been burst?
   (3) What change has occurred in the food substance?
   (4) What changes of shape and size have occurred in the parts originally in the seed?
   (5) Have any new parts appeared?
   (6) Does hypocotyl or plumule develop most rapidly? Why?
   (7) What directions do hypocotyl and plumule take in development, relatively to : —

(*a*) The position of the seed as planted?

(*b*) Any feature of the environment?

(8) What are the relative positions of main and side roots?

Answer by a fully labelled sketch of a typical specimen, and, where necessary, concisely in words.

11. Study in the same manner the germinating Horse Bean.

12. Study in the same manner the germinating Morning-glory.

13. Study in the same manner the germinating Corn.

14. Select any one of the above kinds of seeds, and make a series of outline drawings to illustrate Exercise 10 (7).

15. Your studies (Exercises 10 (7) and 14) have shown you that the positions taken by hypocotyl and plumule in growth are entirely independent of the position of the seed from which they came. Their up-and-down position suggests that gravitation may have something to do with it. To test this, the logical plan is to place two sets of seeds under conditions precisely alike, except that gravitation is allowed to act upon one set and not upon the other. Since, however, nothing upon the earth can be removed from the influence of gravitation, it is necessary so to arrange

N

one set that gravitation may be made to neutralize its own effects. This has been done in
Experiment 1, in which the two sets of seeds
are under the same conditions of temperature,
light, moisture, food supply, and differ only in
their relation to gravitation.

a. Has gravitation anything to do with the position taken by hypocotyl and plumule in
their growth from the seed? Answer by
observation of Experiment 1.

All records of experiments should bring out
clearly : —

(1) Object of the experiment.
(2) Method and appliances used.
(3) Exact results observed.
(4) Conclusions.

b. In precisely what way does gravitation act
to influence the up-and-down position?

c. Express synoptically your conclusions upon
these points : —

(1) For what good do stems grow up and
roots grow down?
(2) By what influence are they guided
into those positions?
(3) By what mechanical means are they
brought around into those positions?
(4) Why is gravitation thus used instead
of other external influences?

Materials. — The seeds may best be germinated in wooden boxes in chopped Sphagnum moss; a greenhouse at a day temperature of 70° is best, and without bottom heat, which makes the roots too slender. These seeds can also be grown easily in the Wardian case (see page 85). Lima Bean and Morning-glory grow faster than the other two, which must be planted two or three days earlier. Six to eight days will bring them into good condition; the best state is that in which the hypocotyl and root are about one to two inches long. The

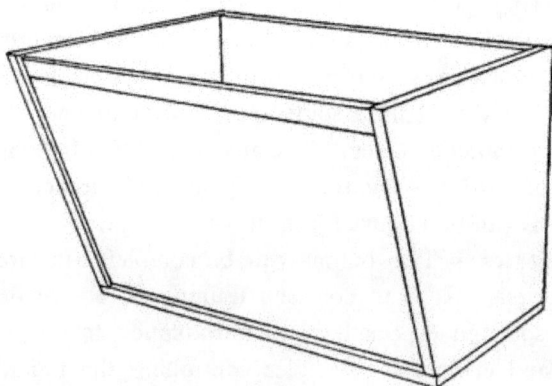

FIG. 16. — Box with sloping glass front for germination of seeds. × ¼.

germinating seeds should, of course, be given alive and growing to the students; hence they should be planted in many small boxes, one to as few students as possible. After many trials I have adopted the following plan : wooden boxes are used, of the shape and size shown in Fig. 16, eight inches long by six wide and five deep, painted for preservation, and with one sloping side of glass slipped into a groove and protected above by a strip of wood; the four kinds of seeds are planted in chopped Sphagnum, in as different positions as possible, eight of each kind in each box. The Lima Beans are planted against the glass, and, growing against it in their descent, show the positions and mode of branching of the roots most beautifully. One of these boxes is supplied to each student, who uses about half of the material, and has it again for the work of the next week, the box in the meantime being returned to the greenhouse, and the seedlings grown on as far as possible. The boxes, if made in quantity at a box factory, cost complete about 12 cents each, and of course can be used many years. The value of supplying to each student his own set of living and growing specimens in a box so arranged as to show the under-ground as well as above-ground parts, is very great, and amply worth the cost and trouble. If this system is not used, the teacher should have some seeds grown in one such box for all the students to see. The advantage of using the moss instead of earth is obvious; it is lighter and cleaner, and the specimens can be removed from it without injury.

Pedagogics. — This outline can be completed in three two-hour periods. It is to continue training in observation and comparison, but especially is an introduction to the morphological and ecological principles controlling the unfolding of the seed into the adult plant, and (most important of all) is an introduction to the nature of irritability, which in plants answers to sensation in animals.

The students should not fail to notice that the root, the root hairs, the turning green of parts exposed to light, the axillary buds of the cotyledons in the brown bean, and the partial disappearance of food substance are new features. They should especially see that it is the elongation of the hypocotyl which raises the cotyledons in Lima Bean and Morning-glory, while it does not increase at all in length in the Horse Bean and Corn. The root is, of course, a new structure developed from the lower end of the hypocotyl, and its beginning is usually marked by a slight constriction or by the first side roots. Students will tend to call the main root hypocotyl, and to call only the side branches "roots," which must be corrected. The structure of the root, including the tips and root hairs, is very plainly seen through the glass, especially by use of a lens, and should be well worked out. Full labelling, to bring out the homologous parts, is very important.

In facts of ecology, they will notice that root grows faster than plumule (of course because absorption of moisture is a first need), and that size of seed, position in which planted, amount of moisture, all have something to do with the different rates of development of the same kinds of seeds, to which some students will probably add a real difference in their living matter, which is strictly true. We have here an introduction to facts of individual variation, so important in evolution. From some of each of the kinds in the boxes, the young plumules should be pinched off, the results to be noted the next week.

It would be of interest also in this connection to study the germination of Horse-chestnut, but practically it is very difficult to germinate.

They will, of course, readily notice in Exercise 10 (7) that the position taken by hypocotyl and plumule (or rather epicotyl), in growth, bears no relation whatsoever to the position of the seed, but that, regardless of this, all hypocotyls bearing the roots grow down, and all plumules grow up. They should then be led to ask what determines this up-and-down position (that is, how does the young plant know which is up, and which is down), whether darkness below, or the moisture, or something else ; they may be encouraged to experiment upon these, and then their minds will be in condition to appreciate the results of Exercise 15.

Other topics of interest and value on germination are : —

How the seed-coats are burst in different seeds.

How the embryo breaks out of the ground.

How the embryos fasten themselves to the ground to give
a resistance to enable the hypocotyl to bore into the
ground.

The behavior of the food substance in germination.

Experiment No. 1. — This experiment is of the utmost importance, since it gives a logical understanding of the true nature of
geotropism, a typical form of irritability, and one of the easiest
to understand. If geotropism is once understood, it will make
all other forms of irritability easily comprehended. Irritability in plants answers to sensation in animals, and a clear conception of it is essential to the understanding of the most
important peculiarities of plant form, movements, and adaptation of the individual to its environment. I believe that one
of the greatest advances that could be made toward placing
the teaching of Botany upon a truly scientific basis would be
through the introduction into it of a correct teaching of irritability. Of course the teacher must first be trained, or train
himself, in this vital subject.

Experiment 1 can be performed very satisfactorily, as follows : Pin to each of two corks, five inches in extreme diameter
and one inch thickness, five or six soaked Horse Beans in as
different positions as possible, though alike on the two corks.
Slip the Beans out to the heads of the pins a half inch or more
from the corks. Place around, over, and under them clean,
moist, chopped Sphagnum moss, and then fit over the corks
thin crystallizing dishes (see Fig. 17), about two and a quarter
inches deep, and wide enough to just hold well on the bevelled
edges of the corks when these are pushed into them (*i.e.* four

and a half inches in diameter). Set one cork upright in a
fixed position; push the horizontal rod of a clinostat into a
hole previously made in the middle of the other, so it will be
kept revolving slowly in a vertical plane. The Beans on the
fixed cork will, in three or four days, all have roots
turned downward, while on the revolving cork they are at
any and all angles, but usually somewhat in the direction they
have in the seeds. It is well to keep them covered from light

FIG. 17. — A simple clinostat. × ¼.

by black paper, but while under observation by the students
the moss may be removed and the glass cleaned and replaced.
Using the Horse Beans, which are the best known to me for
such purposes, this experiment is, with me, always highly
satisfactory. The crystallizing dishes may be omitted, in which
case occasional watering is needed, and the moss must be tied
to the corks, or aluminum dishes could be used.

    Unfortunately, the only clinostats on the market are expen-
sive. Wortmann's is the best; it costs duty-free about $50, and

must be imported from Germany by one of the dealers in laboratory supplies (see page 93). It is useful for many purposes, however, and is a profitable investment. I have made a fair substitute, as follows : buy a Seth Thomas eight-day clock, cost about $5 ; have a watchmaker alter the wheels so that the spindle of the minute-hand will make a revolution in about fifteen minutes (rendered necessary by the shortness of " reaction time ") ; this can be done by shortening the hair-spring and removing each alternate tooth from the escapement wheel ; let him make a brass disk two inches in diameter, with holes on its edge, and an arm to slip over the spindle, so that the disk will revolve parallel to the face of the clock. To this disk may then be fastened, by tacks through the holes, a cork bearing the seeds, as recommended above (Fig. 17). Such a clinostat will not carry large flower-pots, but seeds and seedlings grown in moss show the *principle* of irritability as well as potted plants.

From observation of this experiment, in which the two sets of seeds are under precisely the same external conditions in every respect except in their relations to gravitation, students should be able to deduce the fact that gravitation is a determining influence in giving the up-and-down position to the developing plumules and roots. Since, however, parts grow up against gravitation as well as down with it, it cannot act simply as "weight," but can only serve as a guide or index of direction. The teacher will have to explain that the movement of the growing parts around into the vertical position is brought about by one-sided growth, a subject very easily illustrated by experiment. Gradually the students may be brought to recognize in the process the three elements : (1) an hereditary knowledge (as it were) in the embryo of the advantageousness of sending stems up and roots down ; (2) a power of perceiving from the pull of gravitation, which is up and which is down ; and (3) the use of processes of growth in such a way as to bring the parts around into the up-and-down position. A fair simile helping to make the process clear is that of a sailor starting to cross the ocean, steering by compass. Here, too, are the three elements : (1) the sailor knows to what port it is to his interest to go ; (2) he perceives by observation of the compass which is the proper direction for him to take ; (3) he so adjusts his mechanism of rudder and steam, or sails, as to take him to his destination. The plant uses the pull of gravitation as the sailor uses the compass, purely as an index to direction, and gravitation no more pulls the plant into the up-and-down position than the compass pulls the sailor north and south. In both cases it is previous experience which gives the knowledge of the proper direction to be taken ; in both cases there is use of a guide to show which is the direction ; and in both cases there is a motive mechanism to carry them into the advantageous position. Later studies will prove to the students that gravitation is not only used as a guide to the up-and-down direction, but also as a guide to lateral directions, as in lateral roots and stems, and in many creeping stems and climbing roots ; here, too, the analogy with the compass holds, for the sailor need not go north or south as the needle points, but at any angle between which it is his interest to take, and the compass guides him as well east or west, though it points north and south ; and this is true, also, of gravitation with the plant. The reason why gravitation is used as a guide instead of light (by the stems), or moisture, etc. (by the roots), which also would guide those parts into the proper directions is, no doubt, this, that gravitation acts in the proper direction, with constant intensity, and at all times, while all of the other influences vary in direction, and are sometimes altogether absent.

In geotropism there are many additional experiments very easy to try, and very instructive. For example, while the side roots are growing, the germination box may be tipped up through 45° in a vertical plane, when a beautiful response to the changed direction takes place; to show that the upward growth of stem is geotropic, a small plant may be placed on its side in darkness for a day; also plant a Bean in a small pot of Sphagnum moss, and after it is well up, turn pot and all upside down, and support it in that position for two or three days, after which the moss is to be removed. A most valuable experiment, proving that growth is concerned in bringing roots into the vertical position, is as follows: — Place a soaked Horse Bean in moist Sphagnum moss, with its hypocotyl pointing downward; after the root has grown one inch long, remove it, and, keeping it from drying in the process, mark rings a millimetre apart upon it, from end to end, with waterproof India ink. This may best be done by a thread moistened with the ink and kept stretched on a spring made of wire. Replace the seed in the moss with the root horizontal, and, after it has again turned downward, note the position of the rings.

If students are not satisfied that gravitation is the guide to direction in ordinary plants, but think that it may be moisture which guides the roots into the soil, or light which guides the stems upward, etc., very simple experiments may be invented to prove that these influences do not thus act. Thus, light may be thrown upward upon a plant, turned upside down, by means of a mirror, the plant being covered with a dark box above. Again, seeds may be placed in the centre of a large box of Sphagnum, and watered from above; and other experiments equally simple may be devised by the teacher to meet each point.

V. The Structure and Development of the Seedling

16. Study, at every step in comparison with your records of the earlier stages, the seedling of the Lima Bean.

    (1) Into what has each part of the original embryo developed?

    (2) Are there any new parts not originally in the embryo?

    (3) How are the new leaves placed relatively to the cotyledons and to one another?

    (4) How do the later leaves differ from the earlier?

    (5) How many buds are there, and where are they?

    (6) Where does hypocotyl end and root begin?

    (7) Is there any regularity about the arrangement of new roots as there is about new leaves?

Answer by an annoted sketch, bringing out the above points. The labelling should express clearly the morphology.

17. After the same manner study the Horse Bean seedling.

> (1) Why do the cotyledons remain below ground in this Bean, and rise above it in the Lima Bean?
>
> (2) What effect is produced by this difference upon the growth of the hypocotyls?
>
> (3) Where is the terminal bud which continues the growth of the stem?

18. After the same manner study the Morning-glory seedling.

    Why does the plumule develop so late?

19. After the same manner study the Corn seedling. From what parts do the upper roots come? Is there anything similar in adult Corn plants?

20. From your observations deduce the morphological nature of hypocotyl, cotyledon, and plumule. Express in a sentence.

21. Construct a series of four generalized diagrams of the seedlings studied, expressing in colors (identical with those used in Exercise 7) the comparative morphology of the four seedlings, in comparison with one another and with the seeds from which they grew.

    (Place if possible on the same page with those of Exercise 7.)

22. You have observed in your boxes of seedlings the turning of the plants toward the lightest side; and this turning toward the light is very

well known to you in house plants in windows. The constancy of this turning suggests that light-direction must determine it. To test this, the influence of one-sided light must be removed. This may be done either by placing the plant in the dark (which, however, introduces abnormal conditions), or else by making it revolve so that one-sided light is made to neutralize its own effects. The latter has been done in Experiment 2.

(1) Is light-direction a determinant of bending of green leaves and stems? Answer by observation of Experiment 2.

(2) Is the process of turning (called heliotropism) analogous to geotropism?

(3) For what good do leaves and stems turn toward the light?

(4) Do leaves and stems behave alike as to the positions they take relatively to light-direction?

Materials. — Either seedlings remaining in the boxes of last week, or, since they can hardly grow enough in a week, others grown on in ordinary boxes; they are most useful if the third and fourth leaves show. It is well to grow some of them in the Wardian case, so the students can watch their development.

Pedagogics. — This exercise (needing at least three two-hour periods) is for further training in observation and morphology, but it is especially for the observation of ecological and physiological phenomena, and the use of experiment in their interpretation.

In observation, they should not fail of themselves to see and record, in addition to the more obvious features, the axillary buds of the cotyledons of the Beans, stipules on the Lima Bean (united in pairs at the first leaves), the arrangement of the earlier roots in four ranks (answering to their origin from the four fibro-vascular bundles), and the fact that leaf veins taper from base to tip, and are all united with one another. The position of the terminal bud in the Horse Bean should also be seen correctly.

Exercise 20 is most important to compel clearness in morphological ideas, as is particularly Exercise 21.

In ecology they may be led to see that the failure of cotyledons to come above ground in two of the seedlings is due to their lack of usefulness as leaves on account of their shape. In the Morning-glory, the small supply of nourishment in the seed explains the late appearance of the plumule; the material to make it must first be formed in the green cotyledons. Most students can recall the roots from joints above ground in adult Corn plants. They should be encouraged always to call to their aid any previous knowledge of this kind which they may possess.

Experiment No. 2. — In preparation for this the teacher will do well to direct the students' attention beforehand to the obvious cases of turning toward light in the boxes of seedlings, and the cases known to all of them in house plants. Two simple and similar plants in small pots should be taken (*Tropæolum*, i.e. "Nasturtium," is very good). They should be placed in strong one-sided light, but one of them should be kept revolving in a horizontal plane on a clinostat. Of course if a Wortmann or other large clinostat is available, plants of any size may be used; but if only the small clinostat made from a clock, as recommended for geotropism experiments (page 184), is at hand, then a very light flower-pot, preferably not over three inches in diameter, is needed, and seedlings growing in Sphagnum moss may be used.

Numerous supplementary experiments may be tried, such as allowing the parts to become turned to light, and then exactly reversing them by turning the pot through 180°. Indeed, this simple experiment is almost as satisfactory as the clinostat. Again, the negative heliotropism of roots may very easily be illustrated by the familiar experiment given in most books (as in MacDougal's Physiology, page 59, Fig. 54).

Observation of their experiments and of other cases should lead students to see for themselves that stems turn into the line of the light, while leaves turn at right angles to it, and they can easily be led to see the meaning of this: the light being necessary to the leaves, they expose their flat surfaces to it, while the stems take that direction to help expose the surfaces of the leaves. The very close analogy of the process with geotropism should be emphasized.

VI. The Differentiated Higher Plant

23. Study the Bean plant, a well-differentiated plant. Observe every constant external feature, and properly record.

    Finally, remove it from the pot, wash away the soil, and observe the structure of the roots.

24. In comparison with the Bean, observe fully the Coleus and the Balsam (*Impatiens*). (It is not

necessary to make full records for them.) By comparison of the three plants determine, and express in words : —

(1) Have leaf and bud any constant relationship of position ?

(2) In what positions do flowers (or fruit) originate ?

(3) Do leaves and stem increase in size in some special part, or through their entire extent ?

(4) Is there any regularity in the position of origin of leaves on the stem ? (Invent simple and logical diagrams to show the leaf arrangement in the three plants.)

(5) In what positions do new roots originate ?

25. Diagram the geotropism of the Coleus shoot. In one diagram express the ideal arrangement, and in another its disturbance by light from one side.

26. In growth, such as you have been studying, very important physiological processes occur. From common observation and experience, which may be tested by simple experiments, every one knows that warmth and moisture are necessary to all stages of growth, including germination of seeds. Particularly important, in growth as in many other physiological processes, is its rela-

tion to the two gases, oxygen and carbon dioxide.
This relation may best be investigated in the
germination of seeds, since there it is least
complicated by other processes.

(1) Is oxygen necessary for the germination
of seeds? This can be answered by
an experiment in which a comparison
may be made between one set of
seeds supplied with oxygen, and another set deprived of it. This has
been done in Experiment 3, in which
oxygen is left in the two tubes containing the clear liquids, and absorbed
by pyrogallic acid and potash in the
other. Minutely observe this.

(2) When one of these gases is absorbed,
the other is usually given off.

Is carbon dioxide given off in growth?
This may be answered by an experiment in which a liquid capable of
absorbing carbon dioxide (such as
caustic potash) is so placed that it
will rise in a tube as it absorbs that
gas from a closed space. This has
been done in Experiment 3 in one of
the tubes containing the clear liquid.
The third tube contains simply water.

(3) Is this process like anything in animals?

o

(4) What is the primary meaning and use
of this process in the plant? Has
it the same meaning in the animal?
Your record of (1) and (2) above is to
be worked out as in Exercise 15 (*a*).

27. Prepare a synoptical essay (not over three hundred
words) upon the Germination of the Seed and
the Growth of the Embryo into the Adult Plant.
To be handed in (here the date).

Materials. — Bush Beans (*Phaseolus vulgaris*, var. Golden
Wax) are very easily grown, one in a pot, and may be brought
into flower and fruit in about six weeks. Lima and Horse
Beans grow so large they are unmanageable. Of course other
plants may be used, but the advantage of following some one
kind of plant through its entire cycle is very great, and the
Bean shows a particularly large number of important features.
One plant will do for several students, though the ideal is one
to a student. It is easy to obtain others from florists for
comparison, and Coleus and Balsam (*Impatiens sultani*) are
particularly good, though others will do; one or two of each
of these would be enough for a class.

Pedagogics. — This exercise, in addition to training as before, is intended to give a clear idea of the morphological
composition of the higher plant, and also of the nature of the
process of respiration. As to observation, having reached this
stage, the student should be able to work out and show the constant features of fairly complex structures fully and correctly,
and to represent them well. He should not miss the pulvinus
of the leaves nor the stipels, nor the very important nodules on
the roots, nor the calyx and bracts on the fruit. In morphol-

ogy he should note that flowers (and fruits) originate where buds do; that new parts come either from terminal or axillary buds and that buds produce stems bearing leaves as lateral outgrowths. If allowed to follow his own observations to their conclusions, and not forced into seeing what is not there, he will find that the plant, so far from consisting of a series of joints (phytomera) springing one out of another, grows from continually advancing buds which put out the leaves laterally and branch in the axils of these. The teacher will do well to introduce here an illustrated account of the mode of origin of leaves from vegetative points. The appearance of nodes and internodes is thus not of primary importance, but is incidental; properly, leaves do not stand at nodes; nodes are places where leaves stand. The student can make out also that the stem must grow through a considerable part of its length, but most actively near its tip, and that leaves grow all through their structure. He should also recognize that the root is a single profusely branching structure, originating from a stem. The ecology of leaf, stem, root, should, of course, be fully explained. This exercise affords also a very good introduction to phyllotaxy.

The geotropism of the Coleus shoot may be diagrammed in simple outline figures.

The nodules on the roots of Beans, and their part in supplying additional nitrogen to the plant, should be explained; it is a most interesting and important topic, fully treated in the newer books.

Most important are the facts shown by Experiment 3. It proves the absorption of oxygen and elimination of carbon dioxide by plants in growth, a process identical with that occurring in animals; and it should be the more clearly emphasized here, since there is a general misunderstanding of the process, due to a confusion of it with the gas exchange in photosynthesis. There is no place in the cycle of the plant's life in which respiration can be studied so free from complication with other processes as in germinating seeds.

Experiment No. 3. — Prepare three U-tubes and three upright test-tubes, as shown in Fig. 18, or their equivalents. Half fill the test-tubes with, respectively, water, a strong solution of caustic potash, and a concentrated mixture of caustic potash and pyrogallic acid. Place in one arm of each U-tube a half dozen soaked oats, beneath which is a small wad of

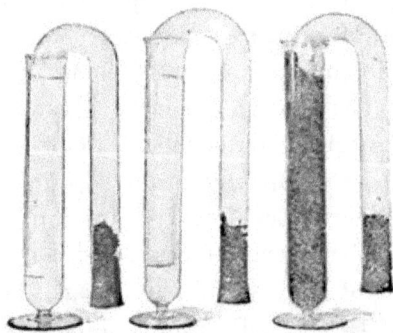

moist Sphagnum, and cork tightly with rubber stoppers; do not allow the arm to become wet above the seeds, or the potash will diffuse over and kill them. Place the uncorked ends of the U-tubes in the test-tubes. The pyrogallic solution will, in a short time, rise in the U-tube about one-fifth of its length, through the absorption of the oxygen; the seeds will not germinate, or, if at all, extremely little.

FIG. 18. — Apparatus for study of respiration in germinating seeds. The tubes contain, respectively, water, solution of caustic potash, concentrated mixture of caustic potash and pyrogallic acid. × ¼.

In the potash tube the liquid will rise to the same height, but more slowly, and the seeds will germinate and grow considerably. In the water tube the liquid will scarcely rise at all, though the seeds will grow as in the preceding. Of course, in the second tube, in the respiration accompanying their growth, the seeds absorb the oxygen and give off carbon dioxide, which is absorbed by the potash, and the latter rises to occupy the space thus left. Some of the brighter pupils, observing the pyrogallic and the potash tubes alone, would say that all the potash tube proves is that something is absorbed, doubtless oxygen, from the air by the plants, and that nothing is proved to be given off since removal of the oxygen alone necessitates the rise of the liquid. In answer to this is the water tube, the failure of the liquid in which to rise proves that something is given off as well as absorbed, and since the only gas absorbed by potash is carbon dioxide, that gas must be given off in volume equal to the oxygen absorbed. This experiment proves that oxygen is necessary to growth, that carbon dioxide is given off, and that the volumes of gas thus exchanged are equal. This exchange is, of course, respiration, necessary to supply energy for growth.

This experiment always works very well with me ; if it fails, it will probably be found that the pyrogallic-potash mixture is not concentrated enough. Practically, it is best to place the pyrogallic acid in the arm of the U-tube, and the potash in the test-tube, when the mixing occurs where it will do most good.

There are other very simple experiments to substantiate these results. Thus the giving off of carbon dioxide may be proved by placing in a closed bottle a number of soaked seeds with a small dish of clear lime water, the milkiness of which, after two or three days, will prove the presence of the gas ; while in a similar bottle without the seeds it will remain clear.

The students will themselves suggest the identity of this process with respiration in animals ; it is also called respiration in plants. The teacher, by a lecture or otherwise, can make it plain that it is by oxidation of food that both plants and animals obtain the energy needed in the work of growth. Of course, the results of this experiment, with a drawing of the apparatus, should be carefully worked out by the students.

VII. Plasticity of the Shoot and Root in Form
and Size

28. In the Bean, Coleus, and Balsam you have studied
three fairly typical Mesophytes. The most
typical or average form would show a vertical,
independent, cylindrical stem, growing by vegetative points at its tip and in the axils of the leaves,
and continued into a main root at its lower end.
The leaves would stand in definite positions
(producing the nodes), separated by spaces (the
internodes), and would be arranged either in
whorls of two or more at a node, or with but
one at a node, and forming a spiral. The leaves
would be simple, thin, horizontal, toothed, ovate
in form, with a petiole and two stipules. But
such a condition is an ideal one and not realized
in any single plant, and very wide deviations
from it occur in plants as result of adaptation
to special habits.

In what respects are the plants on the tables modified from the typical or average condition, and
what is the probable ecological meaning of the
modification?

Your record, of eight plants, should bring out
the name of the plant and scale of the drawing. It is not necessary always to draw the
entire plant, but only its peculiar features.

a. How may simple be distinguished from compound leaves?

b. Do leaves always continue to grow straight out from their points of origin?

29. You have noticed that seedlings turn green as they come into the light, and further observation shows that in general only parts exposed to light are green. It is known to physiologists that starch, which is the real food of plants, is made only in green parts. Is there any connection between these facts, *i.e.* is light essential to starch formation? Answer by Experiment 4.

The exchange of gases, particularly O and CO_2, in these processes, is important. What gas is given off in starch making (photosynthesis or assimilation)? Answer by Experiment 5.

If the gas proven by Experiment 5 is given off, inferentially the other is absorbed, and hence necessary to the process. Is it necessary? Answer by Experiment 6.

Materials. — These should be preferably potted plants, studied in a greenhouse or brought to the laboratory; they need not be injured. Or herbarium material, collected for the purpose on the principle earlier discussed (page 106), may be used. Extreme modifications in adaptation to special function may best be left for another week; here should come "stemless" plants (primroses, houseleeks), flat-stemmed

(Muehlenbeckia), or extremely elongated (climbers), and others with compound leaves of different kinds, some lacking petioles and stipules, and others showing the leading systems of phyllotaxy well, with some to show how this is disregarded in the final arrangement of the leaf blades. Some showing the leaf function assumed by the stem may be added, including the extreme case in the common "smilax" of the greenhouses. In such work as this, the use of a good scientific greenhouse is particularly advantageous. Valuable plants for this purpose, found in such houses, are *Ruscus hypoglossum*, *Colletia*, *Ulex*, *Acacia*, the latter showing the phyllodes, often with compound lower leaves.

Several students may work upon one plant, and they are to be exchanged occasionally. Red labels may be placed upon those to be studied in a greenhouse.

Pedagogics. — This exercise is for training in morphology; also to give an idea of different modes of venation and compounding of leaves, and the main systems of phyllotaxy; but especially it is to make plain how great may be the changes in size and shape of parts while they retain their original nature, a subject of the utmost importance, and at the very foundation of morphology. The students thus trace out how the leaves alter shape, become compound, have or have not petiole and stipules; how the stem lengthens or shortens; how buds multiply or are suppressed, etc., while the relative positions of the parts remain unchanged. Every part may be a centre of variation in form and size. A good conception to place before the students in the summary of their studies is this : to imagine each part indefinitely elastic and compressible, so that any of them may be either greatly drawn out or reduced, while the relationship of position of stem, leaf, and axillary bud remains unchanged.

Care must be taken not to exaggerate the importance of the node. It is not really a distinct structure which incidentally produces a leaf, but it is the place where the leaf stands and hence the fibro-vascular bundles of the stem branch and anastamose, giving the " joint " appearance it often, but by no means always, presents. The " phytomer " has really no morphological existence, as I have elsewhere pointed out (page 149), but is only an incidental result of the way the stem is built. Moreover, this exercise should make plain how readily the stem assumes the function of the leaf, and how little distinct these two are from one another. Hence the plant is best described as made up, not of leaf, stem, and root, but of shoot and root, while the former is further differentiated into stem and leaf, and the leaf may be yet further specialized into blade, petiole, stipules. This relation may be expressed as follows : —

$$\text{Higher Plant} \begin{cases} \text{Shoot} \begin{cases} \text{Stem} \\ \text{Leaf} \begin{cases} \text{Blade} \\ \text{Petiole} \\ \text{Stipules} \end{cases} \end{cases} \\ \text{Root} \end{cases}$$

(see, also, page 149). The main thing now is to teach the fact of the existence of the different kinds of margin, shape, etc., and to show how easily these are derivable one from another, and to give some idea of their meaning. Leaf shape may be treated in a lecture or demonstration somewhat after this manner. The two extremes of shape possible are the circle (accompanying fullest exposure to light) and the line (where crowded), and between them are all variations of ellipses, etc. When an intermediate form is borne out on a long petiole, however, more material is condensed near the petioles, and it gives forms like the ovate, etc. ; when they are crowded together on a short stem so that they would shade

one another, or when without petiole, the material is more
condensed toward the tip, giving the obovate, etc. Leaflets
may be told from leaves by absence of axillary buds, and by
their not originating in whorls nor spirally. There is much
value, it is true, in drawing and naming the different shapes
of leaves, but it is of much the same nature as the fitting
together of some kinds of puzzles; and the same time and
labor may be spent much more profitably upon doing work
which is distinctively botanical and scientific. Still, if the
teacher values terminology as a discipline, here is the place
for it.

The teacher should note that the systems of phyllotaxy
described in the books and expressed by fractions unques-
tionably exist, and may be traced; and a certain amount of
this should be done, enough to give the pupil a clear idea of
its principles; but the teacher should carefully avoid leading
the pupils to imagine they find certain fractions which theoretically ought to be present, for the systems are very easily
thrown out by twisting of the stem in growth or by injuries.
Of course, the phyllotaxy has very little to do with the ultimate position of the blade; it holds true only for the origin
of the leaves in the bud.

The students will be able to do but little with the ecological
explanation of the variations of shape, etc., and here the
teacher must give assistance when he can. It is better to
call attention to such questions, even if they cannot be solved,
than to omit them altogether.

In making observations upon the plants, the students should
read over with each the account of the typical plant given
under 28; it is needful for them to have their attention
directed to each point, or they will miss important features.

Experiment No. 4 is the well-known, valuable, and easy

experiment described in all books for demonstrating photo-synthesis. Select a living potted plant with large, clear, green leaves; keep it two nights and a day in darkness (to empty leaves of starch), then bring into bright sunlight, covering one leaf above and below with tinfoil, in the upper fold of which a figure or letter has been cut; expose this all day to bright light; at evening drop this leaf into nearly boiling water for five minutes (to kill it and swell starch), then place it in strong alcohol warmed over a water bath, which will take out the green in a few minutes and leave it white, or it may simply be left in alcohol until next day. Then place it in a solution of iodine (made by dissolving a little potassic iodide in water and adding solid iodine until it is of a dark wine color), which turns starch dark blue. The letter or other mark exposed to light will stand out dark blue on a white ground. This may be varied in many ways, as described in different books.

Experiment No. 5 is rather difficult to demonstrate well, and the only practicable method is that of collecting in an inverted test-tube over water the bubbles from cut shoots of Anacharis, Cabomba, or some other water plant, which rise and displace the water, as described in all works on physi-ology. The gas must then be tested, which may be done by transferring the test-tube to a very small vessel (slipped under it) and inserting into it caustic potash, when the rise of the liquid will show how much of the gas is carbon dioxide (a very small quantity). If, now, enough pyrogallic acid is added to make with the potash a concentrated solu-tion, the further rise will show how much oxygen (really nearly all of the remainder) is present. This test is difficult to apply, but it is more certain than the usual lighted splinter, or phosphorus, test.

Fig. 19. — Apparatus for study of need of carbon dioxide in photosynthesis. × ⅓.

Experiment No. 6. — Place two similar plants in bell-jars having ground-glass bottoms, which can be sealed with vaseline to ground-glass plates, as in Fig. 19.   In the saucer in one and in the tube in its cork place soda-lime (an absorber of CO_2), and in the other place simply sawdust, in order to have all conditions alike in both except for the absence of CO_2, and its presence, respectively. After about two days' exposure to bright light, the application of the iodine test to leaves will show no starch in those with soda-lime, and abundance in the other, proving that CO_2 is essential to photosynthesis.

VIII. Special Morphology and Ecology of Shoot and Root

The plasticity of root and shoot (leaf and stem) in form and structure is far greater than is shown by the examples studied by you under Exercise 28, and even allows of their modification into special new organs for carrying on new functions; to this end they may be so altered in shape, size, color and texture, as to disguise completely their original nature.   Their positions relative to one another in the plant usually remain unchanged, however, and this forms the best guide to the identity of the disguised parts.

30. In the ten plants selected, what is the exact morphology and probable ecology of the specialized structures they show?

In each case your record should bring out clearly (in the drawings when possible) : —

(a) The evidence which proves their morphology.

(b) Reasons for your view of their ecology.

It will aid in the interpretation of doubtful structures if you will recall the parts or members a typical Mesophyte possesses. The drawings need not include the entire plants, but only the special structures and their connection with other parts.

31. In a sentence explain the idea you attach to the word "morphology"; also to "ecology"; and the exact relationship between them.

Materials. — Living plants from a greenhouse are used, or specimens from the structural herbarium (see page 106, invaluable for this work when living materials are not available) or from the museum, showing highly specialized parts, — spines, tendrils, pitchers, tubers, etc. A good list of such plants is given in Gray's "Structural Botany," Chapter III.

If at the right season, many plants of the native flora are obtainable for this exercise.

Much use can well be made here of good figures, such, particularly, as those in Kerner and Oliver's "Natural History of Plants," and in Schimper's "Pflanzengeographie."

Pedagogics. — This is one of the most valuable of all exercises. It is one of the very best for training the morphological instinct and also for giving knowledge of ecology.

The students should be able in nearly all cases, using relative position as the main guide, to work out with certainty the exact morphological origin of each part, whether from root, stipule, etc. It will be impressed upon them how little the shape, size, color, etc., of organs has to do with their morphology. Of course a complete knowledge of the morphology involves an understanding of the exact steps by which the new organ has been formed, *i.e.* in the case of a pitcher, whether the leaf has infolded and united its edges to form the cup, or (as is actually the case) whether it has grown up as a cup from the start. It will be well for the teacher to have some one or two series of specimens illustrating all the intermediate stages of a particular structure, such, for example, as a Barberry spine. In some cases the student will be able to see what the intermediate steps must have been; but in others this is impossible without a study of embryology, and here (as in the case of pitchers, for instance) it will be necessary for the teacher to supply hints and some information, which students will be prepared to appreciate and utilize after their minds have been once at work upon the problem. It should be made plain to them that the root, leaf, stem, etc., back to which they reduce everything, are not in themselves irresolvable elements, but simply adaptive structures traceable back to still simpler origins, *i.e.* back to the thallus.

On ecology of the structures they can do little better than guess at uses; for, removed from their native homes, the plants can give no idea of their habits. Here is where the outdoor study of native plants through field excursions is most valuable. In ordinary temperate climates the ecological adaptations are so much less marked than in tropical and desert plants, that it will be necessary to use some of the latter in order to give anything like an adequate view of the subject. The teacher must then supply data as to their habits, describing the characters of the desert, the tropical jungles, etc., illustrating by photographs as fully as possible. The teacher must carefully guard against dogmatism in ecology; at the best this division of the science is at present in a very new and undifferentiated state, and even among specialists much of it is but guesswork. A complete study of this subject involves also an examination of the texture, or tissues; for adaptation shows itself in minutiæ as well as in large features, in the suppression of some tissues and excessive development of others; but this work is hardly practicable in an elementary course, except very superficially.

In this connection the teacher should give fully illustrated lectures or talks upon the very important and interesting subject of the ecological groups of plants, — the Mesophytes, Halophytes, etc. These groups may be classified thus, following Warming : —

A. Groups in adaptation to physical conditions.
> 1. Mesophytes, Normal Plants.
>   (Trophophytes, those with winter defoliation. *Schimper.*)
> 2. Xerophytes, Desert Plants.
> 3. Halophytes, Strand Plants.
> 4. Hydrophytes, Water Plants.

B. Groups in adaptation to other organisms.
> 5. Climbers.   9. Insectivora.
> 6. Epiphytes.   10. Myrmecophila.
> 7. Saprophytes.   11. Symbionta.
> 8. Parasites.

If materials are available, here is the place for simple physiologic-ecologic experiments upon such topics as the sensitiveness of tendrils to contact, operation of Drosera leaves, etc.

IX. The Morphology and Ecology of Winter Buds

In climates where a winter stops growth, the living buds must be protected over that time.   How is this accomplished?

32. Study the Horse-chestnut twigs, particularly the buds. Recall your knowledge of how the buds of this tree develop in the spring.

    (1) What markings does the twig show? What is the meaning of each?

    (2) What positions have the buds, and why?

    (3) What sizes have the buds, and why?

    (4) What shapes have the buds, and why?

    (5) What colors have the buds, and why?

    (6) What is the exact structure of the buds?

    (7) What is the morphological nature of each part?

    (8) What is the function of each part?

    (9) What structures have the buds in common with unprotected summer buds, and what accessory to their protection over winter?

Your record should express most of these facts in an annotated drawing, the remainder in notes.

33. Study similarly the Tulip-tree twig.

34. Study also the others supplied.

Outside of the laboratory, examine as large a series of twigs and buds as possible.

35. Prepare a synoptical essay (not over three hundred words) on the General Morphology and Ecology of the Higher Plant.

Materials. — These are abundant everywhere; in place of Horse-chestnut (the best I know) any tree with very large terminal buds will do, especially if containing a flower cluster, as Walnut, Hickory. The bud-scales of the Tulip-tree are modified stipules, hence giving a fine problem in morphology; Magnolia is the same; Beech has the same but less plainly. But any large buds of shrubs are good. For comparison, some unprotected buds of greenhouse plants are needed.

Pedagogics. — This is one of the most useful and satisfactory of all botanical exercises. The objects are large, fairly definite, and the pupil has data enough to enable him to discover for himself the meaning of nearly every feature of structure and ecology. It is particularly good for training in observation and in morphological reasoning, and in relation of structure to use (ecology). It is most important to recall to the students the general habit and mode of growth of the Horse-chestnut, helping by suggestions when memory fails, and leading one member of the class to aid another, until it has been well worked out.

Following are features they should work out themselves : —
Under Exercise 32 (1), the lenticels (whose function as

P

openings for respiration answering to stomata will need to be explained to them, after they have tried to think of a use) ; the leaf scars, with fibro-vascular bundles showing in number answering to the number of the leaflets; rings of bud-scale scars, with a year's growth between the sets; and the old scars of fallen flower clusters.

Under (2), the buds are terminal and axillary.

Under (3), largest buds are toward tip, because the terminal has a flower-cluster, others not, and others are larger toward tip because there is more room there and more light for leaves later ; lower are dormant, and even buried in bark; ask whether every leaf scar has bud in axil.

Under (4), the shape is necessary to hold the many long leaves folded up as compactly as possible.

Under (5), brown, because there is no reason for bright color, and the bud scales take the color of composition of cork which happens to be brown.

Under (6), dissection of a whole bud is needed, and drawings of a bud laid open, and of a vertical section, — or else of individual leaves, scales, etc., and a flower cluster.

Under (7), leaf origin of bud scales is shown by their phyllotaxy, their anatomy, and sometimes by transitions to normal leaves ; really they are the petioles, not entire leaves ; in Tulip-tree, they may readily be discovered to be stipules, — a beautiful case of clear morphology which all should be made to work out. The wool is an epidermal outgrowth from leaves.

Under (8), the scales form a protecting box ; resin prevents rain from soaking through between them ; the wool does not keep out cold altogether, but it prevents injurious suddenness in changes of temperature.

Under (9), the vegetative point with young leaves is in common with the others ; scales and wool are additional.

X. The Minute Anatomy of Root and Shoot

36. In the Balsam, after observing the features of
    the gross anatomy, study carefully the minute
    anatomy of the shoot and root.
    I. The epidermal or protective system.
        (1) Is it continuous and uniform over the
            entire plant?
        (2) Is it removable from the underlying
            tissues?
        (3) Is it smooth or has it appendages?
        (4) Do you find stomata or any equivalent
            for them?
        (5) Is there any green in the epidermis?
    II. The cortical or starch-making system.
        (1) Is it continuous over the entire plant?
        (2) Is it evenly distributed, and, if not,
            where is the green most intense?
    III. The fibro-vascular or conducting and strength-
         ening system.
         (1) Is it continuous through the entire plant?
             Place a spray in the red liquid to aid
             in tracing its course.
         (2) In what order are the bundles arranged
             in the stem?
         (3) How are they arranged in the petioles?
         (4) How are they arranged in the leaf?
         (5) How do they end in the leaf?

IV. The storage system.

All of these systems are to be worked out with simple lens and scalpel.

37. In the young woody stem, what systems may be distinguished?

38. Construct diagrams showing by colors the distribution of tissues in the plant through shoot and root.

Materials. — Balsam (*Impatiens sultani*) is easy to raise, and very good for this use because of its translucent stem, which renders the fibro-vascular system very distinct, though the distribution of its green tissue in the stem is not as sharply differentiated as usual. Coleus is also very good, and almost any herbaceous plant will do. For Exercise 37 any young woody twigs are good, but those with a greenish bark are best.

Pedagogics. — One of the most useful of exercises upon an important phase of anatomy (*i.e.* the contact of, and transition from, the visible to the invisible) commonly overlooked. It is extremely good for training in minute observation, and also as knowledge, for it gives a good comprehensive idea of the distribution of tissues and of the relation of invisible to visible features likely to be missed in an exclusively microscopic study. Far more of minute anatomy can be traced out with the hand lens than is commonly supposed. It also gives, far better than a microscopic study, an understanding of the general physiological uses of the different tissue systems. For best work on this subject the students should previously in some demonstration or lecture have had their attention called to the general physiological conditions which plants must take account of, — protection against drying up, against animal

enemies, exposure of much green tissue to light for starch
making, aeration of the interior cells to allow them to breathe,
conduction of raw sap to the leaf and of the food substances
away, strength to resist winds and other strains, etc. With
all these needs and functions fresh in mind, the students
should be set to work to find out how they are arranged for
in the plant.

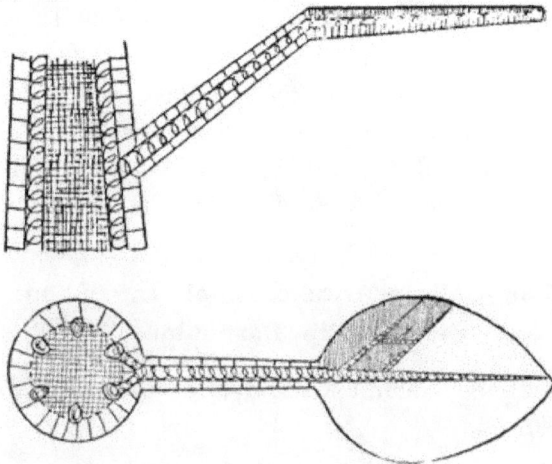

FIG. 20. — Diagram of distribution of tissues in a typical shoot, upper in longitudinal section, lower in cross. Outer line = epidermal system; radiating
lines = cortical system; crossed lines = storage system; spiral lines = fibro-
vascular system. On these systems see page 219.

Important points to be brought out, with their reasons, are :
the lenticels on the stem (which are the successors, structurally
and physiologically, of the stomata of the younger tissues) ;
the greater intensity of the green on the upper, *i.e.* the best
lighted, surface of the leaf; the branching of the bundles at the
nodes, and the running of one branch into the leaf and of
another up the stem ; the fact that the bundles form a ring in
the stem (note the cambium, which, with the vegetative points,

forms a growth system) and that one, two, or three run out
through the petiole and branch profusely, ending either as very
small veinlets anastomosing, or else each ending abruptly in a
small green area (shows well in Asarum) ; the tapering of the
veins regularly and for mechanical reasons. With eosin or
safranin prepare tumblers filled with red dye and place cut
shoots in them ; in a few minutes the fibro-vascular system will
be completely stained. Slides and covers should be given to
allow students to mount all sections in water. Excellent thin
sections can be made with their scalpels, which they may
sharpen on the laboratory whetstone provided for the purpose.
A diagram like that called for in Exercise 38 is shown in
Fig. 20, where the colors are represented by special shading.

XI. The Cellular Anatomy of the Shoot — the Leaf in Particular

In studying cellular anatomy, one is dealing directly
with *cells*.

39. What is the structure of a typical plant cell?

    Answer by a study of the living cell in the
    stamen-hair of Tradescantia. For this the
    compound microscope is needed, the use of
    which will be explained to you.

40. What is the cellular structure of the protective
    system of a typical leaf?

    Answer by a study of the epidermis of Trades-
    cantia, which may be peeled off after a study
    of it in position.

    Notice particularly the guard cells and stomata.

41. What is the cellular structure of a typical leaf?
    (1) What is the structure of the starch-making system?
    (2) Of the conducting and strengthening system?
    (3) Of the aeration system?
    (4) Of the protective system?

Answer by a study of the Rubber-plant (*Ficus elastica*) leaf. Observe carefully the characters of the leaf as a whole; then cut thin cross-sections with your scalpels and compare with the prepared sections.

All of the systems should be represented in a single drawing.

42. It is a well-known fact that leaves give off into the air considerable quantities of water. It is desirable to measure exactly how great the quantity is for an ordinary plant under normal conditions, and also how the rate of this giving-off, or transpiration, is affected by changes in the external conditions. To determine it, the most exact method is weighing, and to employ this, it is necessary to use a potted plant in which all evaporation is prevented, except that through the leaves and stem. This has been done in Experiment 7.

    (1) What amount of water may be given off
        by a plant under normal conditions,

and how is the rate affected by different external influences? Answer
by Experiment 7.

(2) What structures in the leaf are concerned in this process of transpiration?

3) What is the use of transpiration to the
plant?

Materials. — For study of living plant cells, the best object
known to me is the stamen-hair of *Tradescantia virginica*,
which is easily obtained in gardens in late spring and summer,
but not at other times, unless the plants are cut back in the
spring, when they may be made to flower in the late fall; and
if covered at night by a frame and sash, they may be kept in
good condition until near December 1st. *T. pilosa*, common
in greenhouses, gives hairs less excellent but serviceable. The
hairs should be placed in water on a slide under a cover glass.
Another classic object for the purpose is Nitella (or Chara),
which may be found in streams in summer and kept in aquaria
all winter, but they are far less typical than Tradescantia.
The latter is particularly valuable because it shows not only
a typical cell of the higher plants reduced to about the
lowest terms, *i.e.* nucleus, cytoplasm, vacuoles, and wall, but
also shows the cytoplasm in active circulatory movement.
Its simple structure makes it very good to begin with, for in
studying other cells later the student has little or nothing to
unlearn, since others are mostly like it with but additional parts.
Good living plant cells may be obtained also from many epidermal hairs. Of course, knowledge of the cell should be
broadened by observation of mounted as well as other living

cells, and by the study of good figures from books. For epidermis the best object is the *Tradescantia pilosa* or Wandering Jew, common in greenhouses, particularly the leaves with a purple color on the under side. By holding these up to the light the stomata (guard cells) may be seen with a lens, showing green against the purple, and the epidermis easily strips off; it should be placed in water under a cover glass.

For the entire internal anatomy, leaves of India Rubber Plant (*Ficus elastica*) are very good, and are easy to section fairly well with scalpels; they should be cut across the bundles which run out from the midrib. But prepared and mounted microtome sections are necessary for the full demonstration of cellular anatomy. Such sections show the tissues in great completeness and beauty. The Ficus leaf cannot well be used for the epidermis and stomata, for in it these are far from typical.

Pedagogics. — This exercise is to teach the structure of the plant cell, the use of the microscope, and the nature of the cellular anatomy of the higher plants, and also is for training in observation of minute but definite objects. The subject is difficult for beginners, but is altogether too important to be omitted from a well-proportioned course.

In teaching the use of the microscope, its proper function as simply an aid to vision, and not as a tool with mysterious properties of its own, should be made plain from the start. This is best done by leading students to see all possible with the naked eye; — when the limit of this is reached, then simple lenses are to be used, — and when limit of these is reached then low powers of the microscope; certainly all possible of cellular anatomy should first be brought out without the microscope, and they should not take to its use until it is unavoidable. To learn how to focus, move objects, etc., the low power and bits of printed paper on slides are good. Every needful operation of placing the stamen-hairs on the slides, peeling epidermis, etc., should be done by the students; and when sections must be cut for them on the microtome, they should have sections of their own also before them, so they may know just what the former represent.

Naturally, along with Exercise 39 a great deal of description of function, etc., must be given, and from the start the fact that the cell is essentially the protoplasm, and the wall but a passive skeleton or box, should be made plain, as also that the movement in Tradescantia is unusually rapid. Parts they can see are, — the circulating cytoplasm, full of food granules, the nucleus, particularly important in reproduction, the vacuoles, filled with cell sap, forming a reservoir of water and dissolved food substances, and the containing wall.

Under Exercise 40, the precise relation of guard cells to surrounding cells is important as a point of observation; also that nuclei can be seen in epidermal cells, also that guard cells contain chlorophyll. This is particularly good as an exercise in observation. Students must be led to interpret the tissues in terms of cells; i.e. protoplasmic masses with walls, from which, later, protoplasm may be withdrawn.

Under Exercise 41, the drawings should show relation of structures as magnified to those not; i.e. it is better to have students learn to represent just what place their most highly magnified section has in the leaf, as shown on a piece of the leaf unmagnified. Very important is the tracing of the aeration (intercellular) system, its continuity through the leaf, and its connection with the stomata. The palisade layers should be represented as cells — not as a shaded green layer. All of the tissues should be studied from the point of view of function, as a basis for which their knowledge of the conditions of weather, strain, light, physiological work of the leaf, should be recalled to them and further illustrated and explained. The tissues should be studied as protective system, starch-making, strengthening, conducting, aeration, etc., to which the names epidermis, parenchyma, sclerenchyma, ducts, sieve tubes, and intercellular spaces may later be added. In the drawings the cells must be represented as complete structures individually, not simply by uniform shading. Students should be led to view the leaf, not as a mass of cells put together, but as a mass of living substance separated into cells, flattened to expose chlorophyll to light, and needing protection against drying up, a strengthening framework, two sets of conducting tubes, exposure of all living cells to oxygen for respiration, etc. Important, too, is the mode of combination of tissues, — how they are arranged to interfere as little as possible with one another's function.

Something of the excretion system may be made out in the fine large crystal cells (cystoliths) in this leaf.

This work will require at least ten hours from the average student, and should have more. It is worth it.

Naturally, they should be shown other leaves, and especially the mode of ending of the fibro-vascular bundles in the areas of green tissue, which is perfectly plain to the naked eye in Asarum leaves, and with a lens in Cabbage. Also forms of trichomes, hairs, etc., which belong with the protective system, should be shown. These should be in the structural herbarium.

The systems of tissues may well be treated as follows : —

     Protective — epidermis, cork.
     Starch-making — cortex.
     Strengthening — sclerenchyma.
     Growth — cambium and vegetative points.

Conducting : { raw materials — ducts.
{ food materials — sieve tubes.

Aeration — intercellular passages and stomata.

Excretion — special crystal cells.

Storage — pith, medullary rays (and cortex in roots).

Experiment No. 7. — Prepare a plant as shown in Fig. 21, *i.e.* place it in a glass jar and cover with dentist's thin sheet

rubber, tightly tied both to jar and to plant, but pierced by a thistle tube which is closed by a piece of the rubber (or cork) that can be removed. All water must then come out through the leaves and stem. Place the plant on the scale pan of a good balance (the Harvard trip-scale, used in elementary courses in physics, is good), and weigh it at intervals. Add water through the tube, carefully weighing the plant before and after, to find the amount added ; it is very essential to the health of the plant not to add too much, and this can be judged either by the amount given off, or by the appearance of the earth in the pot, which should be allowed to become nearly dry between

FIG. 21. — Method of preparing a plant for transpiration experiments by weighing. ×⅓.

each watering. By placing the plant under different conditions of light, heat, etc., the effect of those conditions upon transpiration may be determined. The larger the plant, i.e. the more leaf surface, the better, since the weighings may then be relatively more accurate. Instead of the glass jar the plant may simply be wrapped in rubber, but as this does not permit the earth to be seen, there is danger of giving too much or too little water, to the great detriment of the results of the experiment. There are many other ways described in various books, of measuring the amount of water removed in transpiration, but none are so satisfactory as weighing.

Transpiration is of great importance, both physiologically and ecologically, and should be discussed fully by the teacher. As in other experiments, as much of the work as possible should be done by the students. The fact that it, like photosynthesis, has no equivalent in the animal economy, should be emphasized. Probably its chief use to the plant is to enable it to lift mineral matters from the soil into the leaves where they are needed.

XII. The Cellular Anatomy of the Shoot — the Stem in Particular

43. What is the cellular structure, and what tissue systems are represented in typical stems of the higher plants? Since stems fall, as to their structure, into two distinct types, it is necessary to select representatives of each.

    What is the cellular anatomy of the Corn stem? What is the distribution of its tissue systems?

Your record should show the exact relation between the gross and the cellular anatomy.

44. What is the cellular anatomy of the Aristolochia stem?

Answer as in the preceding.

45. Construct two diagrams showing by colors the homologies of the tissues in the two stems.

46. With the Aristolochia stem compare a piece of Oak wood. What are the homologous parts?

47. The phenomena accompanying growth in size, and how it is affected by external conditions, are best manifest in stems. To study it properly we must provide some method of measuring its amount, and preferably some method that will be self-recording. This is accomplished by the auxanometer used in Experiment 8.

Under ordinary conditions, at what time in the twenty-four hours does an ordinary plant grow most in length? Answer by Experiment 8.

In what way does temperature affect rate of growth?

Materials. — Indian Corn stems under 1 cm. diameter, put in formaline in summer, are needed. *Aristolochia sipho*, the "Dutchman's Pipe," grows over porches in most towns, and is the classic stem for the purpose, showing with the greatest clearness, in comparison with the Corn, the relation between the exogenous and endogenous composition of the stem.

Pedagogics. — Much as in the preceding. For both stems it is best to use prepared and mounted sections, which show with perfect clearness the cellular character of all the systems. These should be given students, of course, only after they have made out all possible with their own rougher hand-made ones. Most of the work must be done with cross-sections, as the longitudinal can be made to show very little unless cut extremely thin; but students, after some instruction, soon learn to recognize the tissues from a single view.

In Corn, as matter of observation, the companion cells at angles of the sieve tubes should not be missed.

In Aristolochia, the sclerenchyma ring, continuous when young, and broken by expansion of stem later (when the stem twines and needs it no longer for support), should be noted. This is a particularly easy stem for tracing the development of the systems of tissues, which may be done by sections at intervals along the stem. From the Aristolochia, through young Oak twigs, the transition may be followed from the distinct bundles of young stems to the woody mass of older stems, in which the separate bundles are lost. Note homologies of the parts of the young twigs and old wood, particularly in the annual rings and the medullary rays, which latter in the Oak form the shining plates sought for in "quartered" Oak. In Aristolochia, note morphology of pith, and of medullary rays as simply in origin the paren-chyma between the bundles. Also note usual identity of starch-making system and cortex. Trace here homology with the leaf, and how the green parenchyma of the leaf answers to cortex. Note how, in the older stem, epidermis is replaced by cork layers.

Exercise 45 is most valuable; it will impress the real relation of the two types of stems; there will be a difficulty with the cortex and pith in Corn, which are not separable, and colors must merge one into the other. Most important is the practice of recognizing the different tissues with naked eye or hand lens; much histology without a microscope, is possible, and important in many features of adaptation.

It will be profitable to examine cross-sections of other young twigs and different kinds of wood.

Experiment No. 8. — For this, some form of auxanometer or growth-measurer, is needed, of which there is a great variety of forms. Only a self-recording form is really useful, and those on the market are very expensive. A fairly satisfactory form can, however, be made as follows, at a cost of about $2 (see Fig. 22). Buy a dollar clock, four inches in diameter, and remove hands, face, and surplus wheels until only the steel spindle, three-quarters of an inch long, stands up above the works. I have turned on a lathe a cylinder of hard wood one foot long and an inch in diameter, in one end of which a hole somewhat more slender than the spindle of the clock is turned, truly centred. The cylinder may now be forced gently down on the spindle, on which it will revolve evenly once an

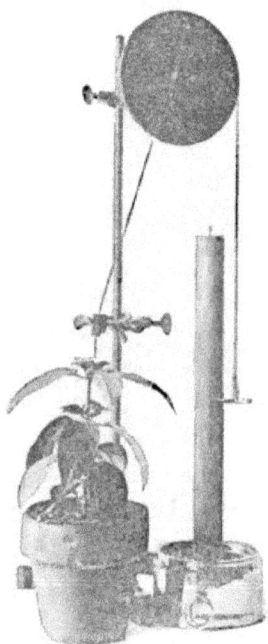

FIG. 22. — A recording auxanometer. × ¼.

hour. Have turned also from good maple a double wheel like that shown in the figure; the outer wheel is grooved, and through the common axis of both a fine, smooth hole is turned by which the wheel will revolve with very little friction on a clean new needle fixed horizontally by solder or sealing-wax to a firm horizontal support. From the very tip of the stem to be studied a fine silk thread, thoroughly waxed to prevent absorption of moisture, is to be run several times around the small wheel to which its end is fastened by a small drop of glue. A similar thread is to be run around the large wheel and fastened, while the free end carries a pen pressing against a paper on the cylinder. This paper should be smooth, put on the cylinder while moistened, and gummed by the free edge, so that when dry it will fit without wrinkles. The pen is to be made from a small piece of slender glass tubing bent into a curve so that both ends, carefully smoothed, rest against the paper; but one of them is drawn into a capillary point and bent and filed so it rests at right angles to the paper. A chronograph ink should be placed in this pen, on which weight enough should be placed to make the wheel turn as the plant grows. As the plant grows the pen will descend, marking a spiral line on the paper, and the distance apart of the spirals where they cross any given vertical line will give the exact amount of growth per hour, magnified, of course, just in proportion to the relative sizes of the wheels. Half-hour periods may be found by ruling two vertical lines at 180° apart, and then removing the paper on one side between them, bringing the vertical lines together. Of course, the paper with the record may be removed from the cylinder for preservation.

Rapidly growing flower stalks of such plants as Hyacinth are very good for this purpose, but any parts that are growing vertically may be used. Variations in temperature can easily be effected where there is a Wardian case; or even by leav-

Q

ing windows open at night, etc. In case there is not room under the wheels for both clock and plant, the latter must be placed to one side, and the thread run over some smooth support, such as a clean screw-eye, as shown in the figure.

Another physiological topic of much importance that may well be taken up here is that of the autonomous movements, particularly circumnutation, of stems. This subject is easy of experiment, and particularly good directions are given by Darwin and Acton in their "Practical Physiology." The glass plate on which records are to be taken may be placed at any desired height by supporting it upon three legs in which grooves to hold the plate are sawn at different heights; the legs are held to the plate by a wire bound around them all. Particularly instructive in this study are the hypocotyls of seeds just bursting from the ground. The well-nigh universal occurrence of circumnutation movements is a point of considerable value as knowledge.

XIII. The Cellular Anatomy of the Root

48. What is the cellular anatomy of a typical root? Answer from a study of the specimen supplied.

49. What is the external structure of the young roots of the Radish?

In particular, what is the structure, distribution, and mode of connection of the root hairs with the root?

How much of the internal structure of the tip can you see with your lenses? (Something more will be shown if you soak the tip for a few minutes in strong potash, then remove and wash it and mount it on a slide in water.)

50. From observation of the appearance of the young roots with their remarkable development of hairs, it would seem probable that these form the structure for absorbing liquids into the plant, and experiments have proven that this is the case. Since, however, observation proves that the hairs and the root tips have no openings, but form a closed system, it is plain that the water must be absorbed through imperforate membranes. The question then arises, Is there any physical process by which liquids can be absorbed through imperforate membranes? This may be answered from Experiment 9, where a membrane (a sort of gigantic hair) has water outside and a solution of sugar inside, precisely as the root hair has. In this experiment, the membrane and an absorbing plant stand side by side.

Can liquids be absorbed through imperforate membranes?

Answer from Experiment 9.

It is also important to know whether the absorption is merely a passive filling of open tubes or an active process that can overcome resistance. This may be learned by attaching a pressure gauge in place of an open tube, as has been done in Experiment 10.

Can pressure be exerted in this process of absorption?

Answer by Experiment 10.

51. Prepare a synoptical essay (of not over three hundred words) on the Cellular Anatomy of the Higher Plant.

Materials. — Roots are much alike in their anatomy, and almost any will do for the anatomy of the shaft called for in Exercise 48; the students' own scalpel sections will show relative development of the principal tissues. Splendid tips and root hairs may be obtained thus: take a small, very porous flower-pot saucer and place in it seeds of Radishes or Mustards, soaked a few hours; cover with another saucer and set it in a dish of water deep enough to keep the seed saucer always wet. In three days the roots and hairs will be perfectly developed. As many saucers and as few students as possible to each is best. It is also well to have a few of the same kinds of seeds sown at the same time in earth, to show how the earth affects the growth of the hairs in comparison with their free and symmetrical development in the saucer. The hairs in the saucers wilt very quickly when exposed to the air.

The Mustard shows the tip and cap, with the radiating lines of growth, even without any special treatment and with a simple lens, but the potash makes them much plainer.

Pedagogics. — An exercise excellent for observation, and particularly valuable for its introduction to the very important subject of absorption of liquids. Under Exercise 49 they should not fail to make out that the zone of hairs advances not bodily, but by growing in front and dying behind. Here also it is profitable to try a simple demonstration experiment proving that the growth of the root is entirely at the tip, which can be done by taking a large root of a germinating Bean and marking it with waterproof India ink at short regular intervals; it is then allowed to grow on farther in a thistle tube, as described in works on physiology. The marks may be put on with a stretched thread dipped in the ink; such marks will not run as do those made with a pen or brush. It is particularly important that students observe that the roots contain no openings, but are a closed system. The distinctness of the growing point, and the protective cap, should be noted.

The explanation of osmosis, the physical process by which the liquids are absorbed in both cases in Experiment 9, is not easy for beginners, nor are botanists and physicists agreed upon the precise nature of the process. It will probably have to be sufficient with beginners to point out to them the physical fact, illustrated fully by Experiment 9, that when a certain solution is on one side of a membrane wettable by water, and water on the other, the water will pass in, while if the membrane is wettable by the solution, some of it will pass out, though not so much as enters; but if the membrane is not wettable by the solution, none of the solution will pass out. For the teacher's own satisfaction, however, the subject should be well worked out, whether he gives it to his students or not. The membranes have no holes that the most powerful microscope can discover, yet there must be openings of some kind, as otherwise the water could not pass. These are supposed to be spaces between the ultimate particles, called "micellæ," of which the membrane is believed to be composed. There are two possible views as to the nature of the osmosis, — one, that water is strongly absorbed by the membrane in virtue of an adhesive attraction between them, but is robbed from the membrane by a stronger adhesion between the dissolved substance and the water, and the limit of the pressure that can be exerted in osmosis would be the limit of this adhesion. The more generally accepted, and probably more nearly correct, view is based upon the fact that the osmotic pressure that can be exerted by any solution is exactly that which the dissolved substance would exert if converted into a gas confined in the same space at the same temperature. Hence the substance is supposed to be in a compressed gas-like condition, constantly exerting expansive pressure, but limited by the boundaries of the liquid in which it is dissolved. Thus the latter is tending always to expand, and hence it will easily absorb any liquid offered to it, as from a wet membrane, and this allows it to expand, and hence to rise in a tube, etc. Therefore, when there is a continuous supply of water, as in our osmometer, there will be a steady rise of the solution until the limit of the gaseous expansion of the dissolved substance has been reached; and hence, also, if this liquid be confined (as in the Pfeffer's artificial cell, particularly well described in Goodale's " Physiology," which the teacher should carefully study in this connection), it will exert pressure upon a gauge. Of course, the energy enabling the gas or dissolved substance to exert its pressure is derived from heat in the atmosphere. The marked difference between the osmometer made by the diffusion shell and the root hair, in that the former allows some of the sugar to pass out, while the latter does not, must be emphasized. In the root hair there is not only a membrane comparable with the parchment, viz. the cellulose wall, but an additional one, a lining film of protoplasm, which in the root hairs (but not always in other living cells) is impervious to the sugar solution. Another difference between them is that the membrane opens into an open tube, while the root hairs do not, but communicate through lines of living cells with the ducts in the root and stem. How the water gets from the hairs unto these ducts is as yet entirely unknown. But the primary physical process is the same in the parchment cup and the living root-hair cell, and in Experiment 9 one may say that on the one side we have a great number of tiny hairs, and on the other a single gigantic one.

Experiment No. 9. — Take two burettes of 16 mm. diameter, and remove the bottom up to 2 cm. below the beginning of the graduation, and smooth the cut end in the flame. Over one of them fit a soaked diffusion shell of 16 mm. diameter (which may be obtained of Eimer and Amend of New York), and tie it to the burette very tightly with a waxed thread. Fill it with a thin solution of molasses up to the zero mark. Such an instrument is a very efficient osmometer. The molasses may be made stronger if quicker working is desired. These shells are the best arrangements for the purpose I know of, but if they are not available, a piece of good

FIG. 23. — Osmometers constructed from a parchment shell and from a living plant. ×⅓.

parchment (generally obtainable of bookbinders) may be soaked and stretched tightly over the lower end of the burette ; of

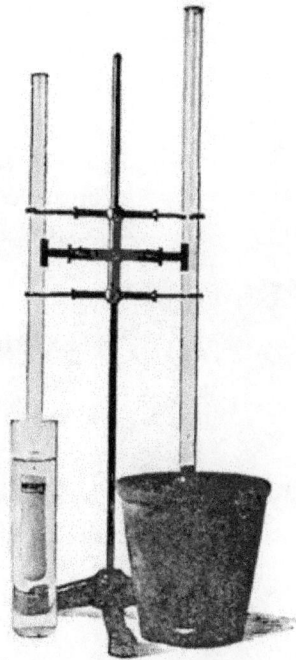

course, with its smaller surface, it works far more slowly than the diffusion shell. To the other burette a plant is to be attached as shown in Fig. 23. Select an actively growing plant with a stem about the size of the inner diameter of the burette. Cut it off an inch from the earth, and attach it to the burette by a rubber tube fitting over both it and the plant, and tie it tightly to the plant with a rubber band so there can be no leak. When plant and osmometer are placed side by side, as in Fig. 23, the demonstration is very instructive. A film of oil may be placed on the liquid in both tubes to prevent loss of any of it by evaporation. Of course, very careful records of the rise should be made by the students. Exact measurement has great pedagogic value in itself, and a habit of preferring precise quantitative results to loose generalizations should always be cultivated.

Experiment No. 10. — A simple and effective pressure gauge may be made as follows (Fig. 24) : Take a glass tube with a glass stop-cock at one end, and graduate it in millimetres and centimetres with India ink applied with a stretched thread. Select a vigorous plant with a stem about the diameter of the inside of the tube, and cut it off an inch from the ground. On the top of the stump fit a short piece of rubber tubing thick enough to make stump and outside of tube the same diameter, and slip another piece of tubing over this and the tube so as to make a water-tight joint. This joint must next be made

FIG. 24. — Gauge for measuring root pressure. × ½.

inexpansible to pressure from within, which can be done by winding it tightly and carefully by several turns of tire-tape (used for repairing bicycle tires). Enough water should then be put into the tube to bring it up to the zero mark, and the stop-cock (a perfectly air-tight one) should then be closed. The water forced out from the stump will then compress the air column, and the exact pressure exerted may be calculated by Mariotte's law, — that pressure is inversely proportional to the volume of the gas. Thus, suppose the air column is compressed to three-fourths of its former length ; this means a pressure upon it of four-thirds. But it had one atmosphere, that is, three-thirds, upon it at the start ; hence the additional pressure exerted by the water will be one-third of an atmosphere, or about five pounds to the square inch. This implies that the readings shall be taken always at the same temperature, which is not difficult to manage, and it neglects a slight error due to the water vapor in the tube, but the latter is at the most very small. The ingenious teacher can make a tube without the stop-cock, perhaps even a closed test-tube, do. The plant is, of course, to be watered regularly, but not too much, or the roots will soon die of suffocation. Another very instructive experiment upon roots is one to show their hydrotropism, a most important irritable property ; methods of demonstrating it are given in all physiological works.

XIV. The Anatomy and Morphology of the Flower

52. What is the structure of the essential parts of the flower — pollen-grain and ovule ?

Answer from a study of the material supplied.

The microscope must be used. After examining
the pollen dry, add water, and observe the effect.
After examining the ovules, placed on a slide
in water, as fully as possible, add potash, which
may make their structure clearer.

53. What is the exact structure of the Scilla flower?

(1) Of what distinct parts is it made up?

(2) In what relative positions are these arranged?

In addition to your drawings, construct diagrams
which shall show in ideal horizontal section the
ground plan of the flower, and in ideal vertical
section, the vertical plan. In each case represent sections through the most typical parts of
the structures. The two diagrams are complementary to each other, and one need not repeat
what the other shows.

54. In your earlier studies you have found that the
flower originates as a branch does, *i.e.* from an
axillary bud.

In what way does the Scilla flower answer to a
branch, *i.e.* in what way have stem and leaves
altered their shapes and positions to form the
parts of the flower?

Represent by diagrams the intermediate stages
between leaves and the parts of the flower.

55. What is the function of each part of the flower?

56. After the same manner study the Hyacinth flower.
Represent by diagrams only.

Materials. — For pollen and ovules, those of Scilla or Hyacinth are good. For the structure of the flower, if this work comes in summer, Trillium or Buttercup are both very good. If in winter, *Scilla siberica*, squill, is the simplest and most typical plant available; it is extremely easy to raise in shallow boxes; the bulbs, each supplying several flowers, are cheap, and any skilful gardener can have them ready on a given date. Tulips are good, but expensive. Next best is Hyacinth, the single white Roman kind, but this is much less simple and typical. These are grown for sale in most greenhouses, and flower so abundantly they are not expensive. But of course others will answer, though kinds with superior ovary must be selected. In summer many simple forms may be collected and preserved in formaline, or even dried and pressed, but in the latter case they must be soaked out in warm water, and are far inferior to fresh flowers. It would be a mistake to give a pressed flower to a pupil to begin with.

Pedagogics. — A study in observation, recording, and knowledge of the flower. The introduction to the flower through the study of pollen and ovule is extremely important as helping to impress upon students what is really essential to it. An account of fertilization and its meaning should here be given. Under Exercise 53 the study is purely in anatomy. At this stage of their work they should be able without special help to work out fully and correctly the structure of such a flower as the Scilla, and to represent it well. They should not miss such points as that three of the perianth parts are outside of the other three, that there are three cells to the ovary, that the ovules are on a central placenta, and that the anthers contain pollen. But too much detail, such as kinds of ovules, dehiscence of anthers, etc., must not be expected at this stage, else time is lost and proportion is destroyed; the

most essential things first, is the best rule. Terms for the
principal parts — perianth, petals, sepals, etc. — and for the
conditions of union of parts, — gamopetalous, gamophyllous
(for parts of a perianth), etc. — should be given after the need
for them has been felt.

The construction of the diagrams is the most important
pedagogical part of this exercise. They will be spoken of
below.

In the morphology, the students should of themselves recognize that receptacle is stem which remains short, that petals
and sepals are leaves; but stamen and pistil, particularly
anthers and ovules, will puzzle them. They should be
allowed, or, if necessary, led to see that the latter are not
homologous with anything they have yet studied; in fact, so
far from representing modified edges of leaves, etc., they are
as distinct from leaf or stem as these are from root, and they
are older than the leaf or the stem (see page 146). They
are sporangia containing spores, an inheritance from the non-
flowering plants, with certain appendages added. The ovule
(nucellus) is a spore-case containing a single spore (macrospore or embryo sac) whose germination produces the egg-
cell, the whole surrounded by one or two protective coats.
The anther is a spore-case containing spores (microspores or
pollen-grains) whose germination produces ultimately the
pollen tube with its contents. The pistil is composed of infolded leaves with the spore-cases on their edges. It is a mistake to try to homologize the ovary, style, and filament, with
blade or petiole of a leaf, for the differentiation into blade and
petiole is an attribute of the foliage leaf only, not of the spore-
bearing leaves, which, it is possible, have not been derived at
all from foliage leaves (see page 147). I have found it in
my own experience most profitable to teach the correct mor-

phology of these parts, including ovule and pollen-grain, from the start; pupils understand it as readily as they do the formal and partly incorrect morphology current in many of our text-books, and they have nothing to unlearn later.

It is usually assumed that a perianth tube, such as the Hyacinth has, is composed of united petals and sepals; this is not strictly true, for the tube is probably not made by the union of the bases of petals and sepals, but by a ring of tissue under the bases of the petals and sepals, which is one continuous structure, a sort of ring leaf, and not six united parts (see page 148). The point is very important for an understanding of the composition of complex flowers.

The function of pollen and ovule can best be given them through an account, fully illustrated by diagrams, of the process of fertilization. That of calyx can be illustrated by reference to buds where it is a protection to the young parts. As to the showy corolla, its use can be brought out by such a line of reasoning as this: Experiments and observation have shown that better seed is produced when pollen and ovule come from different plants; this requires the locomotion of pollen from one plant to another; this is often brought about by wind, but that is a very wasteful method; a much more economical mode of locomotion of the pollen would consist in using some agency which could be made to move from one flower to another; small animals, particularly insects, form such an agency, but some inducement must be provided to make them visit the flowers; this is generally done by nectar, on which they feed; but the place where the nectar is must be shown them so they may find it; this is done either by strong odors, or else by color; the special structure developed to hold the color is the corolla. Later the argument may be continued thus: not only must the insect be brought to the vicinity of the nectar, and

therefore of the pollen, but it must be made to approach the
nectar in such a way as to leave upon the stigma the pollen
it has brought, and to take a new supply; hence the different
shapes and sizes of flowers — shape being chiefly to make the
insect enter the flower in a position proper to secure the
pollination, and size being in general related to the size and
form of the visiting insect. This mode of reasoning must be
used with great caution, and not allowed by the pupils without
the most complete evidence for their arguments. It is impossible, however, for them to work out without great time and
labor the true theory of the flower, and a theoretical account
of it like this is much better than none. It would be far
better to obtain a basis for such a description by study of wild
flowers out of doors in summer.

Like most other teachers, I have used blank forms for
description of flowers, but have abandoned them, not because
they are not valuable if properly used, but because much more
good can be obtained from the same amount of time and
labor spent as here recommended. Besides, the blanks imply
a great amount of work on terminology, which again, while
far from valueless, does not, nevertheless, in my opinion, constitute the best use that can be made of the students' energy
and time.

Of great value in the study of flowers is the representation
of the fundamental facts of their structure by horizontal and
vertical diagrams as called for under Exercise 53. These are
intended to represent, not superficial features of form, etc., so
much as fundamental relations of number, relative position,
coalescence, etc. Ground plans for this purpose are given in
all works upon floral structures, but the equally useful vertical plan is much less used. As an example, there are here
given these diagrams for Scilla and Hyacinth (Figs. 25, 26).

The following principles should be observed in their construction. The two kinds are complementary to one another,
and it is not necessary to try to show in one what is already
brought out in the other. Relations of number, alternation
and coalescence of like parts, are brought out in the horizontal,
and general form and adnation of unlike parts in the vertical.
Conventional signs, as shown by the accompanying examples,
can be used for the parts. Form should be shown only so
far as possible without interfering with the clearness of repre-

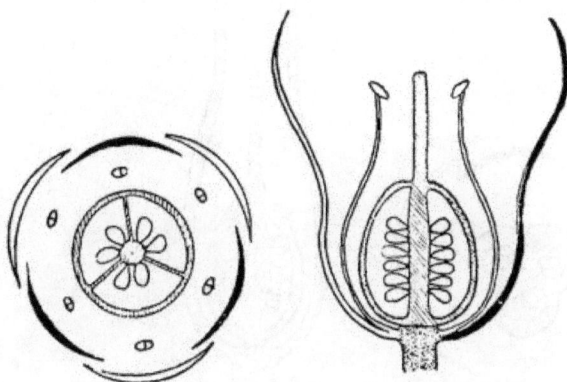

FIG. 25. — Diagrams of Scilla flower. Receptacle dotted; carpels cross-
lined; petals black; sepals and stamens unshaded.

sentation of the more essential features. They should be
constructed with the most rigid exactness, every spot and
line having its meaning, and no confusion of lines allowed.
Particularly important is the insertion of parts upon the receptacle and upon one another; and lines should not be
allowed to touch one another in the diagram except in
order to represent parts grown together in the flower.
The help of compasses, etc., should be required, if necessary,
to make them symmetrical. Teachers should remember, how-

ever, that while these diagrams are extremely useful servants
they are bad masters. In my own experience I have found

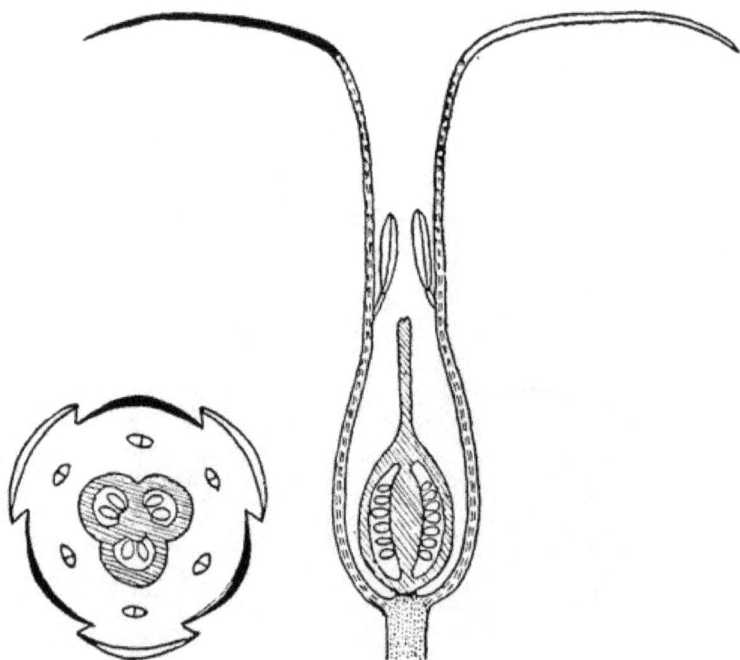

FIG. 26. — Diagram of Hyacinth flower.   The vertical lines show the perianth
tube;   other shading as in Fig. 25.

nothing to equal them for compelling clear ideas on the
part of the student.

XV. The Morphology and Ecology of the Flower

57. What is the exact structure and morphological
    composition of the Snowdrop flower?
    Express in the horizontal and vertical diagrams.

By special shading bring out the exact morphological nature of each part.

What adaptations to cross-pollination does it show?

58. What is the exact structure and morphological
composition of the Narcissus flower?

Answer as for the above.

59. What is the exact morphological composition of the
Primrose flower?

Answer as for the above.

60. Tabulate the resemblances and the differences
between the Scilla and the Primrose.

61. What is the exact morphological composition of
the Fuchsia flower?

62. What is the morphological composition of the
Eupatorium flower?

63. What is the morphological composition of the
Cytisus flower?

In the diagrams the irregularity may in a general way be represented; but it must not be
allowed to interfere with the clearness of the
ground plan.

Materials. — The Snowdrop is the very best of flowers with
inferior ovary, and worth much trouble to obtain. Fuchsia is
also extremely good for a flower with inferior ovary, and easy to
obtain at florists. Narcissus and Freesia are also good. Primroses, grown in large numbers for sale by all greenhouses, are
very good and not expensive. The range of materials used in
the preceding exercise and in this is ample to explain fully the

R

morphology of the flower. For a Composite, which can readily be understood at this stage, the small white Eupatorium often grown in greenhouses is very good, but others will do, such as Cineraria, or *Senecio petasites*. If irregular flowers are added, any papilionaceous flower will do, of which some kinds are always in greenhouses.

Pedagogics. — More specialized flowers are here taken up, and some that are irregular. The question of the morphological composition of the wall of the inferior ovary must be faced. Students may best be introduced to this by stating to them the fact, illustrated by diagrams, that every flower, no matter how specialized, originates as a set of originally distinct leaves on a conical receptacle; let them reason from this in the case of the Snowdrop, and if they are not previously prejudiced by the calyx-adnate-to-the-ovary theory, they will readily see that the stamens, petals, and sepals must stand on the receptacle, which therefore must form the wall of the ovary by growing up in the form of a hollow cup, while the carpels form the roof over it, and also the partitions. This is the morphology which embryology sustains. In the Fuchsia the morphology of the ovary is the same, but here, in addition, a tube is formed after the manner already spoken of for Hyacinth and Primrose In the Composite flower the morphology is very like that of Primrose and Fuchsia, *i.e.* the ovary is a hollowed-out receptacle on the top of which the sepals, often finely divided into a pappus, and the corolla-tube stand.

In diagramming the flowers of the Snowdrop, etc., it is well to use shading of the different parts, to represent their exact morphology. In the preceding diagrams (Figs. 25–26) this is done, and in the same way it may be done in the Fuchsia (Fig. 27). In diagramming the irregular flowers, as the Cytisus, a part of the irregularity can be shown, but it must never be allowed to interfere with the clearness of the ground plan of the flower.

In laboratory study, students cannot do much if any practical work on cross-pollination, but it is well to keep the matter before

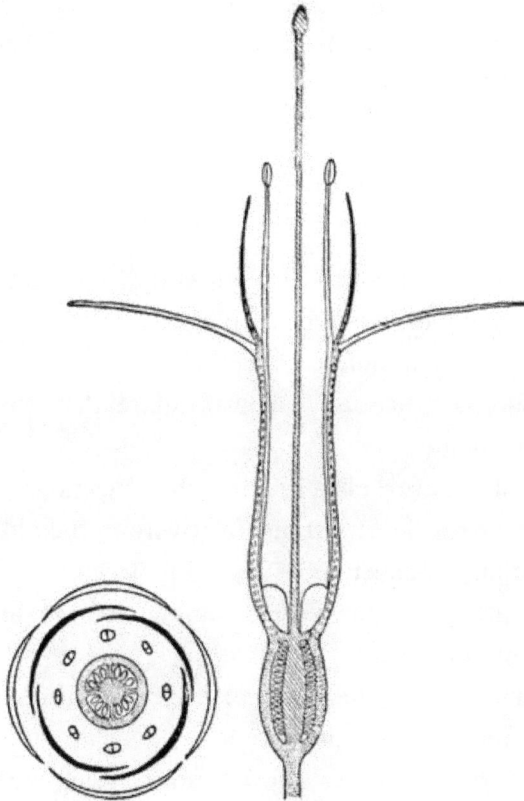

FIG. 27.— Diagrams of Fuchsia flower.   Shading as in Figs. 25 and 26.

their attention, particularly in flowers like the Fuchsia, which shows splendid nectar glands, and in irregular flowers where the alighting-place and definite path of the insect can be traced. Something can be accomplished in the laboratory by imitating with brushes, etc., the operations of the insect in particular flowers. The teacher can find facts on the mode of cross-pollination of many common flowers in Müller's "Fertilization," and with this as a basis can make valuable demonstrations to the class. A lecture or talk, illustrated by diagrams, upon this most interesting of subjects will be appreciated and have great meaning at this stage.

XVI. The Morphology and Ecology of the Flower. — *Continued*

64. In each of the ten flowers supplied, what is the identity of each visible part?

    Answer by annotated sketches.

    Can you trace any special adaptations to cross-pollination?

65. In the flower clusters, in what positions do the younger flowers stand relatively to the older?

    Can any connection be traced between the size of a cluster and the size or number of the blossoms composing it?

    What does a cluster probably mean in connection with cross-pollination?

66. Construct a series of diagrams, using colors, to show the intermediate stages in the development from a simple conical vegetative point of —

    a. A flower with all parts distinct.

    b. A flower with superior ovary, but the other parts united into a tube.

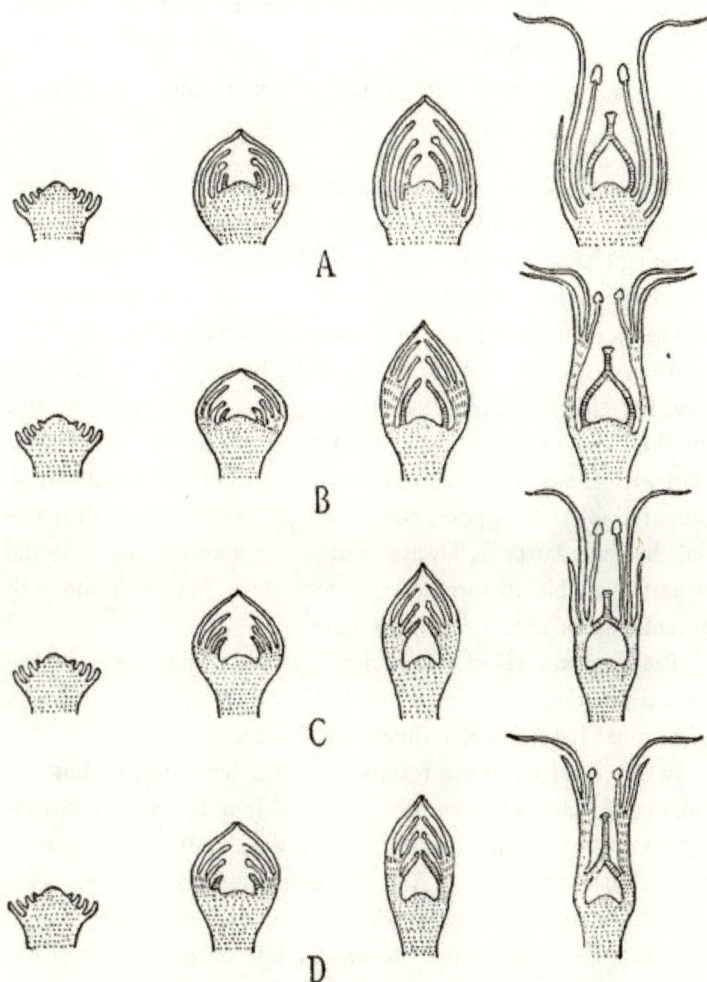

FIG. 28. — Diagrams to illustrate the morphology of typical flowers. *A*, hypogynous; *B*, perigynous; *C*, epigynous; *D*, epigynous with prolonged "calyx tube." Receptacle is dotted; carpels are cross-lined; "perianth tube," or "calyx tube," vertically lined. Sepals, petals, and stamens are unshaded, but may be distinguished by their relative positions.

c. A flower with inferior ovary, but other parts distinct.

d. A flower with inferior ovary, but other parts united.

Materials. — For Exercise 64 there should be obtained from a greenhouse some of the more special forms of flowers, such as Begonia, Calla, Orchids, Poinsettia, Narcissus, etc., of which the different specialized parts are to be recognized by the student, and reduced to their proper categories of sepals, petals, etc. Of course, only a few of each kind can be had, as they are expensive, but they need not be taken apart, or only partially, and by the teacher; they may be passed from one student to another. To some extent herbarium specimens, especially if prepared for the purpose, could be used instead of fresh material, but the latter is always best. For Exercise 65 material is best available in summer, but something may be done with greenhouse or even herbarium material.

Pedagogics. — Under Exercise 64 comes some good morphological practice, training in the habit of recognizing similarity of original nature under diversity of form.

In Exercise 65 some terminology will have its use, but this subject of flower clusters, while of considerable value in classification, is not of much interest otherwise. Of course, care is taken by the teacher to select the more marked types.

Most valuable is the work of Exercise 66. This cannot, it is true, be made from observation, but must be worked out theoretically, as shown on the accompanying diagrams (Fig. 28). The exercise has great value in contributing to ideas of almost mathematical clearness. A student cannot construct these diagrams who does not perfectly understand the morphology of the complex flowers. Series *a* is about like the

Scilla, except that it is supposed to have a distinct calyx and corolla. *b* is not like any of the flowers studied, but is nearest like the Hyacinth except for the distinct calyx and corolla. *c* is like the Snowdrop, and *d* like the Fuchsia. This morphology differs much from that in use in the manuals, but is more nearly correct, as shown by embryological studies.

XVII. The Morphology and Ecology of the Fruit

67. What is the exact structure and morphological composition of the six dry fruits supplied?

>   (1) What has become of each of the parts of the original flower, *i.e.* sepals, petals, stamens, receptacle, ovary, style, and stigma?

>   (2) How are the carpels or receptacle modified and arranged to form this fruit?

>   (3) What is the morphology of the new or accessory parts, — wings, etc.?

>   (4) In what places, morphologically, is the dehiscence?

>   (5) How are the seeds probably scattered?

> Answer by diagrams and drawings as far as possible. Under (2) bring out the leaf or stem homology in each case.

68. What is the exact structure and morphological composition of the six fleshy fruits supplied?

> Answer as for Exercise 67.

69. Prepare a synoptical essay, not to exceed four hundred words, on the Morphology and Ecology of the Flower and Fruit.

Materials. — In part these may be bought in markets, in part must be collected the year before. Typical follicles are Columbine, and Larkspur or Monk's-hood; legumes are green Beans or Peas, or Locust pods; winged fruits are Maple and Elm; others are Poppy, Sunflower, Shepherd's Purse. Of fleshy fruits, good kinds are Grape, Tomato, and Orange (especially navel), Apple, Banana, Cherry (canned are good), Strawberry, Cranberry. Many others can be used, but these are particularly typical and obtainable.

Pedagogics. — This is a very valuable exercise for morphology. The students cannot, of course, from the fruits alone settle all points of morphology, but they can settle many; and as for the rest, it will be a most valuable exercise for them to form their hypotheses, and then have these confirmed or otherwise by the teacher, who will supply missing data. This, under rigid control, is a truly scientific procedure, indeed the greatest help of the investigator. Their interpretation will be greatly aided if they are shown pictures of the flowers from which the fruits come, — that is, if the flowers are not themselves available. The ideal would be for them to have several stages from the flower to the fruit. It will be best to take the fruits up in order, the simplest, *i.e.* the follicle, first, then the legume, and so on.

This must not be made a drawing lesson in still life, at least not in class; it can be largely worked out by diagrams. Diagrams which are halfway between the carpels and unmodified leaves are particularly valuable, but, of course, the fruits should be drawn and labelled for structure also.

It is not worth while to give students unusual terms, such as sarcocarp, etc., which are not used in descriptive works, but follicle, legume, drupe, etc., should, of course, be supplied as they are needed.

The true morphology of the fruit should be taught; *e.g.* in the Apple, the flesh is mainly receptacle, with a little of it from carpel; in the Cranberry, it is receptacle, etc. Particularly important is a clear idea of the Cherry, with part of the carpel forming stone and the other part pulp. Something similar to this separation occurs in the Orange, where the skin is separable; it is a part of the carpels. The pulp of the Orange is a growth of hairs from the inner (upper) faces of the carpellary leaves, though these hairs are not unicellular. The whole subject of the morphology of the pulp is of great interest; it originates in a variety of ways. An account of seed locomotion (a subject always of great interest to students) should be taken up here in much more detail than was possible near the beginning of the course.

It may be noted here, by the way, that the word "ecology," so often used in this work (spelled œcology in the Century Dictionary, and defined there) is coming rapidly into general use to express adaptation of plants and their parts to external conditions.

DIVISION II

THE NATURAL HISTORY OF THE GROUPS OF PLANTS

THIS second part of our course is a study of the Natural History of Plants. It investigates the habits and structure of these organisms, and the relations of their shape, size, color, positions, and cellular texture to their modes of nutrition, growth, reproduction, locomotion, and protection. Whenever possible the plants are to be observed as they grow naturally and undisturbed in their native homes.

I. The Algæ

70. What is the structure and ecology of the Pleurococcus?

Your record should bring out clearly : —

  a. The exact appearance to the naked eye of the organism as it grows under natural conditions, and a description of those conditions. Whatever annotated drawings will not bring out is to be added in notes.

  The exact structure of the organism, not only in two dimensions, but in all three, with expression of the true size.

     c. Both the vegetative and the reproductive parts.

    d. Ecological connection between structure and habits.

71. What is the structure and ecology of Spirogyra? Answer as for Exercise 70.

72. What is the structure and ecology of Fucus? Answer as before.

73. What is the structure and ecology of a typical Red Seaweed?

74. From your own studies, from the specimens and pictures examined, reading, and other sources of information, concisely describe : —

    (1) the range of habitat in Algæ;

    (2) of color;

    (3) of size;

    (4) of shape ;

    (5) of texture.

75. Prepare a concise essay, not over two hundred and fifty words, on the Natural History of the Algæ, emphasizing their ecology.

Materials. — The aim is to give first a typical unicellular form, reproducing by fission, and a filamentous conjugating form next. Zoösporic forms can hardly be used for study of processes of reproduction, because of practical difficulties. As this study is more ecological than systematic, it matters little just what forms are taken as long as they are typical. In summer many Algæ are available, but in winter I have found that Pleurococcus and Spirogyra give the optimum resultant between accessibility and representativeness of their respective groups. Pleurococcus may be found on the damp, shaded bricks and flower-pots of any greenhouses, but, since many other Algæ occur in those places, it is necessary to examine the material carefully; it may be obtained also from the bark of trees, on the damp, shaded side, where, as a green film, it is sufficiently familiar. Protococcus occurs in about the same situations, but, as it reproduces by zoöspores only, which are extremely difficult to demonstrate, it is less useful. As to Spirogyra, it is a classic object, and good for many purposes. Conjugating and zygosporic material must be secured the autumn before (or may be bought from the Cambridge Botanical Supply Company), and, with vegetative material, may be preserved in formaline. But much better is material kept all winter in a dish or tank in a greenhouse, as it can then be seen of its natural color and appearance. In all cases the material alive and on its natural substratum should be brought into the laboratory. Fucus may be collected on the coast in summer and preserved in formaline, or may be obtained alive and fresh at any time of year from the Cambridge Botanical Supply Company on a few days' notice. For its proper study, sections through the conceptacles are needful, and these may be made by the students themselves with a sharp scalpel, the end of the frond being held between two flat pieces of pith. There is no Red Seaweed known to me which is easily obtainable alive in quantity and in condition to show its reproductive parts to students. I have had to use herbarium specimens of various species for the vegetative structure, and to supply the reproductive structures of a typical form from diagrams, using the Kny series for this purpose. The students copy this diagram with explanations; it is not a good principle, but it is better than nothing.

Pedagogics. — Up to the present this course has been concerned chiefly with training in botanical principles, using the higher plants as a basis; information has been subordinate to the cultivation of eye and hand, and to the formation of scientific instincts. From this time on, the object is to lead the student to make a close and sympathetic personal acquaintance with the chief kinds of living plants; information becomes of equal value with training, and the means for acquiring the former is of greater value. It is true that but few kinds can be studied; hence it is best to select forms as representative as possible of the great leading groups. The aim should be, using a thorough study of these as centres, aided by collections, figures, and reading, to secure, through the medium of their own senses, the impression upon the minds of the students of a clear, sharply lined picture of the place in nature of each group, — what kinds of places it lives in, how it obtains its nourishment and reproduces, and the meaning of the most constant characters of form, color, etc., and how each is related to the other groups.

There are so many excellent books upon the natural history of the different groups that extended directions are here unnecessary. These books are referred to in Chapter VII, but particularly practical and valuable to the teacher are Spalding's, Barnes's, and Atkinson's works.

It is of first importance that students see the forms they study growing alive in their native places, and that they look upon them not inertly, but with active curiosity, which will be the case if the teacher keeps properly before them problems to be solved. Next to this, and supplementary to it, is the study of herbarium, or museum materials, photographs, and prints. If it is not possible for them to see the plants alive and at home, then the teacher should describe to them as vividly as possible, and with all available illustrations, just where and under what conditions they grow.

The compound microscope is, of course, necessary from the start.

The representation of the living plant, no matter how small, called for under Exercise 70, *a*, seems to me most important. Even in Pleurococcus, where a single plant cannot be distinguished with the naked eye at all, the student gains far more accurate knowledge of the exact place of the organism in nature if he has to draw and describe the appearance of the colonies or masses of it, than if, after a hasty glance at the living form, he confines his studies to magnified images of it. Throughout this study of natural history of plants I regard this representation of the appearance of the entire organism, as it looks alive, as one of the most important of all exercises. Colored drawings are the best, and the fullest scope should be given the artistic talents of students; but a black and white drawing, with colors, etc., explained in notes, is better than a coarsely or badly colored picture.

Exercise 70, *b*, is also important; it may be brought out by shading, but also, and for most students better, by imaginary cross-sections. These are of great value for testing the students' knowledge.

Exercise 70, *c*, is necessary; they should acquire the habit of seeking for the reproductive parts.

Exercise 70, *d*, can best be answered in a concise paragraph. Of course, for Pleurococcus, this is most simple, as the plant is unicellular, and all functions are performed by one cell; substances are absorbed anywhere over the surface. Spirogyra and other floating forms are but little more complex; such simple forms hardly have any "ecology." In the more complex forms, however, adaptations become more pronounced,

and the thinness and fineness of division of the forms always
immersed, in adaptation to the difficulty of obtaining sufficient
oxygen, the toughness and elasticity and powerful holdfasts of
those dwelling between tide-marks exposed to the full force
of the waves, the bladders for floating, the red and brown
instead of green colors, probably in adaptation to the peculiar
light-conditions, all should receive attention.

Exercise 74 is particularly valuable as calling attention to
the chief elements in adaptation.

This study of Algæ will occupy at least four, and better
six, two-hour periods.

In this kind of study, I think collection is of great value.
The collecting instinct is one of the chief attributes of the
successful naturalist, especially of him who studies whole organisms. The taking, the preparing, the keeping of specimens,
all have value in increasing acquaintance, and the reference
to them from time to time afterward is a pleasure and a
profit. But as most people lack this inclination, it is better
to make the collecting voluntary. Algæ are easy to preserve.
Of Pleurococcus, a little may be scraped carefully off, put on
a small piece of paper, moistened, well spread out, and then
placed between driers, with a bit of cotton cloth over the
alga to keep it from sticking to the upper paper. Spirogyra
should be floated out well in water, then paper should be
slipped under it, and the whole lifted from the water, to be
dried afterward as in the Pleurococcus. These may then be
mounted in the book-herbarium described elsewhere (see page
110). A most valuable series may be made by mounting
specimens of all the plants studied in this Part II, and thus
would result a very instructive collection of types representing
the groups from Algæ to Phanerogams. This would accord
with one division of the plan earlier recommended (page 104).

II. The Fungi

76. What is the structure and ecology of Bacteria?
Answer fully as for Exercise 70.

 On the basis of instruction given you, and your reading, write a paragraph upon the economic aspects of this group of Bacteria.

77. What is the structure and ecology of Yeast?

 Write a paragraph upon the economic aspects of Yeast.

78. What is the structure and ecology of Bread Mould?

79. What is the structure and ecology of the Mushroom?

80. From your own studies, from specimens and pictures examined, reading, and other sources of information, concisely describe : —

 (1) the range of habitat in Fungi ;

 (2) of color ;

 (3) of size ;

 (4) of shape ;

 (5) of texture.

The outlines, here so brief, will of course be made much more detailed by the teacher, to accord with the details of his own instruction. Henceforth in this series, only the advantageous forms for study will be indicated.

Materials. — These are easily obtained. Bacteria from hay placed two or three days in water in a warm place ; or from Lima Beans soaked two or three days, and from many other sources which all of the books tell about; Yeast, of course, from yeast-cake placed overnight in water with a little sugar (Yeast may be made to form spores by growing on plaster of Paris plates as described in books) ; Bread Mould by keeping bread several days moist in a warm place ; Mushrooms from the markets, or canned, may be used. Of course, there are numerous other easily available forms of Fungi that may be used if time allows, but it is most important to use the chief types.

Pedagogics. — As under Algæ. A full account of economics should be given in lectures or demonstration, including the importance of Bacteria not only in decay, diseases, etc., but in the arts (cheese-making, etc.), the nitrification of soils, fixation of nitrogen in Leguminosæ, etc. Full descriptions should be given also of the important forms of Mildews, Rusts, and others of economic importance ; those which are studied in the laboratory will form a good basis for the theoretical study of others, for each form actually studied by the student illuminates many others.

If the material on any particular stage is insufficient, diagrams may be copied ; this is better than skipping an important stage altogether.

It is especially important to keep constantly prominent the *place-in-nature-and-among-other-plants* idea, which requires generalization and active use of the imagination.

Very important is the phylogenetic relationship of one of these groups to another : the teacher should make it plain that Fungi are forms degenerate through parasitism from the Algæ, and not a homogeneous group, but derived from different sub-groups of the Algæ (see Fig. 29). There is great difference of opinion as to the details of the relationships of these groups. There is a valuable discussion of the subject in Campbell's " Evolution of Plants."

s

III. The Lichens

81. What is the structure and ecology of the Parmelia, a typical Lichen?

    Answer as for the earlier groups.

82. From your various sources of information, concisely describe: —

    (1) the range of habitat in Lichens;

    (2) of color;

    (3) of size;

    (4) of shape;

    (5) of texture.

83. Prepare a synoptical essay, of not over three hundred words, on the Natural History of the Fungi and Lichens, emphasizing their ecology.

Materials. — The common Parmelia growing upon trees everywhere is excellent, and both thallus and apothecia may be obtained at all seasons. The students, of course, should collect it for themselves, and describe it as it appears in its native situation.

Pedagogics. — As in the preceding. The group is of great interest on account of the remarkable symbiosis of the Algæ and Fungi, which should all be made plain.

To show properly the two elements of the thallus, and especially to show the spores in the apothecia, students' scalpel sections are not alone sufficient, in which case they may be shown how to section in pith with a razor (described in Strasburger and Hillhouse, "Practical Botany"); it is much better to let them do this sectioning than to have it done for them. Prepared microtome sections are very useful also.

IV. The Bryophytes

84. What is the structure and ecology of a typical Hepatic (Marchantia)?

Answer as for the preceding groups.

85. What is the structure and ecology of the true Moss?

Answer as for the earlier groups.

86. What is the range of habitat, of color, size, shape, texture, in Bryophytes?

87. Prepare a synoptical essay of two hundred words, upon the Natural History of the Bryophytes.

Materials. — *Marchantia polymorpha* is the most available Hepatic, and may be collected in summer with the archegonia and antheridia ripe, and placed in formaline, or may be bought from the botanical supply companies. For a Moss, Polytrichum, or "pigeon-wheat," collected in summer and put in formaline, is good. It is necessary to have the antheridial and archegonial material as well as the nearly ripe spore-cases.

Pedagogics. — It is generally difficult to find the archegonia of any Moss, and probably the teacher will think it best to use mounted preparations for this. Of course, as before, the habitat will be examined or explained, and the students' ideas broadened by an inspection of museum and herbarium material and of pictures. They should particularly be shown some of the simpler floating forms of Hepaticæ, alive if possible, to emphasize the derivation of this group from the Algæ, and also specimens of Anthoceros, which doubtless gave origin to the Pteridophytes. Of course, the position of the Mosses as a side and barren branch, and their alternation of generations, will be emphasized.

V. The Pteridophytes

88. What is the structure and ecology of the Fern plant?

   Does the vegetative structure agree in its general composition with that already studied by you in the flowering plants?

89. What is the structure and ecology of the pro-thallus stage of the Fern?

90. What is the structure and ecology of the Selagi-nella?

91. Prepare a synoptical essay of not over three hundred words, upon the Natural History of Pteridophytes emphasizing their ecology.

Materials. — Any Fern with the sori in good condition will do; material may be obtained from greenhouses at any time. The prothalli are very difficult to find out of doors, but are easy to obtain in abundance on neglected flower-pots, walls, and earth in greenhouses, particularly in badly kept ones. While the general structure of the prothalli is easy to deter-mine, the exact structure of archegonia, in particular, is very difficult for beginners, and it may be needful to have sections prepared for them. Full directions and other very valuable matter on this subject will be found in Atkinson's "Biology of Ferns" (see bibliography on page 137). Selaginella may be obtained from greenhouses, and shows the macrospores and microspores in good condition in winter and early spring.

It is very desirable also to allow the students opportunity to examine at least one typical Club-moss (*Lycopodium*) and an Equisetum.

Pedagogics. — This is an important group from many points of view. Ecologically it is important as the lowest large land group, though its fertilization belongs to a water habit. The alternation of generations, here at its plainest, should be emphasized. The structure of the sporangia may be easily made out fully. In Selaginella, the two kinds of spores may be seen, but it is not possible to get the prothallus, etc., and this must be described from pictures; and if the teacher is skilful, some idea can be given of the nature of heterospory and of its significance in the transition from Cryptogams to Phanerogams, though it must be confessed this is a difficult topic for beginners. The relation of the Pterido-phytes to the Hepaticæ, through Anthoceros, should be empha-sized, and the table of relationships given students as in Fig. 29. It is well to keep these trees of relationship constantly before them, adding each group as it is studied.

VI. The Gymnosperms

92. What is the structure and ecology of the Pine?

   Can you homologize the vegetative structure with that of the higher plants studied earlier by you?

   Can you homologize the reproductive structures with those of the Pteridophytes just studied? Or with the flowering plant formerly studied?

262   THE TEACHING BOTANIST

93. Prepare a synoptical essay, not over two hundred
words, upon the Gymnosperms, emphasizing
their ecology and relationships.

Materials. — Male and female flowers of the Pines may be
obtained in June and kept in formaline; along with the young
stages should be collected some of the year-old cones.

Pedagogics. — This group is important but not easy to study
fully. It is easy enough to study the male flowers of Pine,
in which the homology of the anthers and pollen should be
kept plain by calling them microsporangia and miscrospores.
In the female flowers it is very difficult to make out parts of
the embryo-sac and the egg-cells, and diagrams must be used.
The homologies should be made plain by calling the nucellus
macrosporangium and the embryo-sac *macrospore*, though
strictly it is only the very young and ungerminated embryo-sac
that is macrospore and the later stage they study is its germinated condition. It is not worth while to try to homologize the
scales of the cones with carpels; the homologies are extremely
doubtful, in any case, and the most recent studies seem to show
that the Gymnosperms are not the ancestral forms of the
Angiosperms at all, but have come off from the Pteridophytes
by a distinct branch as shown in the table in Fig. 29. It will
be well to give the students some idea of the Cycads, and
particularly the Cycas with its antherozoids, of which there is
a good account in Atkinson's " Elementary Botany."

VII. The Angiosperms

a. *Monocotyledons*

94. What is the structure and ecology of the forms
supplied?

b. *Apetalous Dicotyledons*

95. What is the structure and ecology of the Willow?
96. What is the structure and ecology of the other Apetalæ supplied?

c. *Polypetalous Dicotyledons*

97. What is the structure and ecology of the forms supplied?

d. *Gamopetalous Dicotyledons*

98. What is the structure and ecology of the forms supplied?
99. Prepare a synoptical essay, not to exceed three hundred words, upon the Angiosperms, emphasizing their ecology and relationships.
100. Prepare a genealogical tree which shall represent the descent of the Angiosperms from the Algæ, showing their probable phylogenetic connections.

Materials. — This part of the course will come in the spring when there is abundance of material out of doors, and no doubt plenty of plants in all of the groups are everywhere available. Of course, also, typical kinds may be kept in formaline, or even dried, though the latter is not to be recommended. The aim should be to secure representatives of the different leading groups.

Pedagogics. — This part of the course is the most familiar to teachers, and hence there is little need for special directions. For an apetalous form, Willow is particularly good, since it so nearly represents a theoretically primitive flower, *i. e.* one consisting of stamens, or pistils, and a nectar gland ; the latter is theoretically the first accessory part of a flower to appear, and the most important part in adaptation to insect visits ; the color develops later to show where the nectar is, and brings with it the need for a color-carrier, which office is assumed by some of the stem leaves, originating the corolla.

There is much doubt as to the position of the Monocotyledons. By some they have been considered as a side branch of Dicotyledons, but the weight of evidence at the present day places them in the ancestral line of the Dicotyledons, and derives them through the low water-plants from the heterosporous higher water-ferns. The division into Apetalæ, Polypetalæ, and Gamopetalæ is not natural, but is convenient from an ecological point of view. The teacher will, of course, keep before the students the part played by adoption of insect pollination in development of the corolla, and of increasing specialization for insect visits in the development of the gamopetalous condition. The teacher should use in classification the Engler and Prantl system employed in the most modern books ; it is much more natural than the Bentham and Hooker system largely employed hitherto. The aim of the teacher in selecting materials for study in this division should be to represent the leading groups and families, as they are given in the best modern books, as, for instance, in Campbell's " Evolution," and in Strasburger's "Text-Book." Of course, at this time naturally comes the use of manuals for identification of species. As I have already said, however, I do not think the time of all members of the class should be taken for this, but extra voluntary classes should be formed of those especially interested in the subject. Of great profit, too, are field excursions, when attention should be given not so much to collecting specimens for identification, but to seeking illustrations of the principles of adaptation they have studied earlier in the course. Practi-

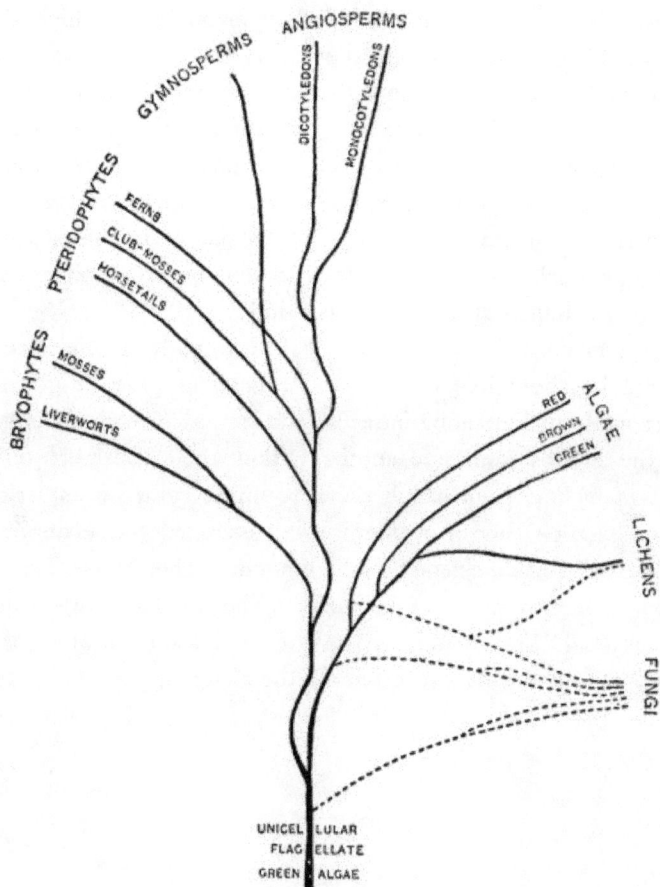

FIG. 29.—Hypothetical tree of relationship and descent of the leading groups of plants.

cally, I have found that large field parties are not profitable, and that ten is a maximum number that should be allowed

to go. At these times materials may be gathered for the
structural herbarium earlier recommended (page 109). When
analysis work is undertaken, it is better done in the field; for
which purpose the students take their manuals (for which the
leather-bound field editions should be used) with them. In
this work every effort should be exerted to make them
acquainted with families as well as with species. Each plant
studied should be made to contribute not only its own name
and systematic position, but should also make firmer and
clearer the students' knowledge of the larger groups, so that
the whole scheme of classification will come to stand out as
a sort of unified structure in his mind.

A genealogical tree, such as is called for under Exercise 100,
would be about like that on the preceding page (Fig. 29).
The mode of branching indicates the supposed mode of origin
of the groups from one another; thus the Algæ were once
the main line, from which the Bryophytes came off as a side
branch, soon, however, themselves assuming the main line,
while the Algæ became a side branch. The Mosses are a
side branch from the Liverworts. The Pteridophytes came
off as a side branch from Bryophytes, but soon took the main
line, and so on, as expressed in the diagram.

INDEX

WORKS ON BOTANY.

ATKINSON (G. F.). — The Study of the Biology of Ferns by the Collodion Method. For Advanced and Collegiate Students. By George F. Atkinson, Ph.B., Associate Professor of Cryptogamic Botany in Cornell University. 8vo. $2.00.

BOWER (F. O.). — Practical Botany for Beginners. By F. O. Bower, D.Sc., F.R.S., Regius Professor of Botany in the University of Glasgow. 16mo. 90 cents.

DARWIN (F.) and E. H. ACTON. — Practical Physiology of Plants. By F. Darwin, M.A., F.R.S., and E. H. Acton, M.A. Crown 8vo. $1.60.

DARWIN (F.) — Elements of Botany. By F. Darwin, M.A., F.R.S. With Illustrations. Crown 8vo. $1.60.

CAMPBELL (D. H.). — The Structure and Development of the Mosses and Ferns (Archegoniatae). By Douglas Houghton Campbell, Ph.D., Professor of Botany in the Leland Stanford, Junior, University. 8vo. $4.50.

MURRAY (G.). — An Introduction to the Study of Seaweeds. By George Murray, F.R.S.E., F.L.S., Keeper of the Department of Botany, British Museum. With 8 Colored Plates and 88 other Illustrations. 12mo. $1.75. *Reg.*

SETCHELL (W. A.). — Laboratory Practice for Beginners in Botany. By William A. Setchell, of the University of California. 12mo. 90 cents.

STRASBURGER (Dr. Edward), DR. FRITZ NOLL, DR. HEINRICH SCHENCK, and Dr. A. F. W. SCHIMPER. — Lehrbuch der Botanik für Hochschulen. With Illustrations. Translated by Dr. H. C. Porter, of the University of Pennsylvania. 8vo. $4.50.

STRASBURGER (E.). — Handbook of Practical Botany. Edited from the German by W. Hillhouse. Third Edition. With numerous Illustrations. 8vo. $2.50.

VINES (S. H.). — A Student's Text-book of Botany. By Sidney H. Vines, M.A., D.Sc., F.R.S. With numerous Illustrations. Cloth. Complete in one volume. $3.75.

WARMING (E.). — A Handbook of Systematic Botany. By Dr. E. Warming, Professor of Botany in the University of Copenhagen. With a Revision of the Fungi by Dr. E. Knoblauch, Karlsruhe. Translated and Edited by M. C. Potter, M.A. With 610 Illustrations. 8vo. $3.75.

THE MACMILLAN COMPANY,
66 FIFTH AVENUE, NEW YORK.